ELEMENTARY CONCEPTS OF

TOPOLOGY

By

PAUL ALEXANDROFF

with a Preface by
DAVID HILBERT

translated by
ALAN E. FARLEY

DOVER PUBLICATIONS, INC.
New York

Copyright © 1961 by Dover Publications, Inc.
All rights reserved under Pan American and International Copyright Conventions.

Published in Canada by General Publishing Company, Ltd., 30 Lesmill Road, Don Mills, Toronto, Ontario.
Published in the United Kingdom by Constable and Company, Ltd., 10 Orange Street, London WC 2.

This Dover edition, first published in 1961, is a new English translation of *Einfachste Grundbegriffe der Topologie,* as published by Julius Springer in 1932.

International Standard Book Number: 0-486-60747-X
Library of Congress Catalog Card Number: 61–1982

Manufactured in the United States of America
Dover Publications, Inc.
180 Varick Street
New York, N. Y. 10014

TRANSLATOR'S PREFACE

IN TRANSLATING this work, I have made no attempt to revise it, but have merely tried to preserve it in its proper historical perspective, and have made only a few minor modifications and corrections in the process. I have inserted a footnote and several parenthetical notes in an effort to clarify the material or to indicate the current terminology, and have actually changed the notation in several places (notably that for the homology groups) to avoid confusion and to bring it into consonance with modern usage.

I wish to express my gratitude to Jon Beck, Basil Gordon and Robert Johnstone for their invaluable aid in the preparation of this translation.

ALAN E. FARLEY

Ann Arbor, Michigan
January, 1960.

PREFACE

FEW BRANCHES of geometry have developed so rapidly and successfully in recent times as topology, and rarely has an initially unpromising branch of a theory turned out to be of such fundamental importance for such a great range of completely different fields as topology. Indeed, today in nearly all branches of analysis and in its far-reaching applications, topological methods are used and topological questions asked.

Such a wide range of applications naturally requires that the conceptual structure be of such precision that the common core of the superficially different questions may be recognized. It is not surprising that such an analysis of fundamental geometrical concepts must rob them to a large extent of their immediate intuitiveness—so much the more, when in the application to other fields, as in the geometry of our surrounding space, an extension to arbitrary dimensions becomes necessary.

While I have attempted in my *Anschauliche Geometrie* to consider spatial perception, here it will be shown how many of these concepts may be extended and sharpened and thus, how the foundation may be given for a new, self-contained theory of a much extended concept of space. Nevertheless, the fact that again and again vital intuition has been the driving force, even in the case of all of these theories, forms a glowing example of the harmony between intuition and thought.

Thus the following book is to be greeted as a welcome complement to my *Anschauliche Geometrie* on the side of topological systematization; may it win new friends for the science of geometry.

DAVID HILBERT

Göttingen,
June, 1932.

FOREWORD

THIS LITTLE book is intended for those who desire to obtain an exact idea of at least some of the most important of the fundamental concepts of topology but who are not in a position to undertake a systematic study of this many-sided and sometimes not easily approached science. It was first planned as an appendix to Hilbert's lectures on intuitive geometry, but it has subsequently been extended somewhat and has finally come into the present form.

I have taken pains not to lose touch with elementary intuition even in the most abstract questions, but in doing so I have never given up the full rigor of the definitions. On the other hand, in the many examples I have nearly always dispensed with the proofs and been content with a mere indication of the state of affairs which the example under consideration served to illustrate.

Mindful of this latter end, I have picked out of the extensive subject matter of modern topology only one set of questions, namely those which are concentrated on the concepts of complex, cycle and homology; in doing so I have not shied away from treating these and related questions in the full perspective appropriate to the modern state of topology.

With respect to the basis for the choice of materials appearing here, I have included a paragraph (**46**) at the end of this book.

Of course, one cannot learn topology from these few pages; if however, one gets from them some idea of the nature of topology—at least in one of its most important and applicable parts, and also acquires the desire for further individual study—then my goal will have been reached. From this point of view let me direct those of you who already have the desire to study topology to the book written by Herr Hopf and myself which will soon be printed by the same publisher [see footnote 4—A.E.F.].

I should like to express my warmest thanks to S. Cohn-Vossen and O. Neugebauer, who have read this book in manuscript form as well as in proof and have given me worthwhile advice on many occasions.

My sincere thanks also to Mr. Ephrämowitsch at Moscow and Mr. Singer at Princeton, who most kindly undertook the drawing of the figures.

P. ALEXANDROFF

Kljasma at Moscow,
May 17, 1932.

CONTENTS

INTRODUCTION

1. The specific attraction and in a large part the significance of topology lies in the fact that its most important questions and theorems have an immediate intuitive content and thus teach us in a direct way about space, which appears as the place in which continuous processes occur. As confirmation of this view I would like to begin by adding a few examples[1] to the many known ones.

1. It is impossible to map an n-dimensional cube onto a proper subset of itself by a continuous deformation in which the boundary remains pointwise fixed.

That this seemingly obvious theorem is in reality a very deep one can be seen from the fact that from it follows the invariance of dimension (that is, the theorem that it is impossible to map two coordinate spaces of different dimensions onto one another in a one-to-one and bicontinuous fashion).

The invariance of dimension may also be derived from the following theorem which is among the most beautiful and most intuitive of topological results:

2. *The tiling theorem.* If one covers an n-dimensional cube with finitely many sufficiently small[2] (but otherwise entirely arbitrary) closed sets, then there are necessarily points which belong to at least $n + 1$ of these sets. (On the other hand, there exist arbitrarily fine coverings for which this number $n + 1$ is not exceeded.)

---

[1] One need only think of the simplest fixed-point theorems or of the well-known topological properties of closed surfaces such as are described, for instance, in Hilbert and Cohn-Vossen's *Anschauliche Geometrie*, chapter 6. [Published in English under the title *Geometry and the Imagination* by Chelsea, 1952—A.E.F.].

[2] "Sufficiently small" always means "with a sufficiently small diameter."

For $n = 2$, the theorem asserts that if a country is divided into sufficiently small provinces, there necessarily exist points at which at least
three provinces come together. Here these provinces may have entirely

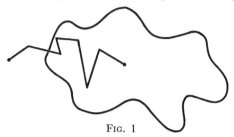

FIG. 1

arbitrary shapes; in particular, they need not even be connected; each one
may consist of several pieces.

Recent topological investigations have shown that the whole nature of
the concept of dimension is hidden in this covering or tiling property, and
thus the tiling theorem has contributed in a significant way to the deepen-
ing of our understanding of space (see **29** ff.).

3. As the third example of an important and yet obvious-sounding
theorem, we may choose the *Jordan curve theorem:* A simple closed curve
(i.e., the topological image of a circle) lying in the plane divides the plane
into precisely two regions and forms their common boundary.

2.   The question which naturally arises now is: What can one say about a
closed Jordan curve in three-dimensional space?

The decomposition of the plane by this closed curve amounts to the fact
that there are pairs of points which have the property that every polygonal
path which connects them (or which is "bounded" by them) necessarily
has points in common with the curve (Fig. 1). Such pairs of points are said
to be separated by the curve or "linked" with it.

In three-dimensional space there are certainly no pairs of points which
are separated by our Jordan curve;[3] but there are closed polygons which

---

[3] Even this fact requires a proof, which is by no means trivial. We can already
see in what a complicated manner a simple closed curve or a simple Jordan arc can
be situated in R^3 from the fact that such curves can have points in common with
all the rays of a bundle of rays: it is sufficient to define a simple Jordan arc in polar
coordinates by the equations

$$\varphi = f_1(t), \qquad \psi = f_2(t), \qquad r = 1 + t$$

are linked with it (Fig. 2) in the natural sense that every piece of surface which is bounded by the polygon necessarily has points in common with the curve. Here the portion of the surface spanned by the polygon need not be simply connected, but may be chosen entirely arbitrarily (Fig. 3).

The Jordan theorem may also be generalized in another way for three-dimensional space: in space there are not only closed curves but also

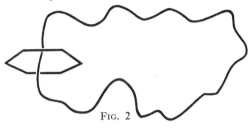

FIG. 2

closed surfaces, and *every such surface* divides the space into two regions— exactly as a closed curve did in the plane.

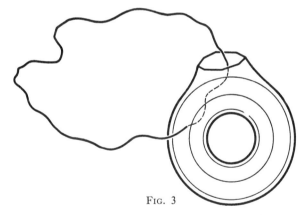

FIG. 3

Supported by analogy, the reader can probably imagine what the relationships are in four-dimensional space: for every closed curve there exists a closed surface linked with it; for every closed three-dimensional

where

$$\varphi = f_1(t), \qquad \psi = f_2(t)$$

define a continuous mapping of the unit interval $0 \leq t \leq 1$ onto the unit sphere $r = 1$.

manifold a pair of points linked with it. These are special cases of the *Alexander duality theorem* to which we shall return.

3. Perhaps the above examples leave the reader with the impression that in topology nothing at all but obvious things are proved; this impression will fade rather quickly as we go on. However, be that as it may, even these "obvious" things are to be taken much more seriously: one can easily give examples of propositions which sound as "obvious" as the Jordan curve theorem, but which may be proved false. Who would believe, for example, that in a plane there are three (four, five, . . . in fact, infinitely many!) simply connected bounded regions which all have the same boundary; or that one can find in three-dimensional space a *simple* Jordan arc (that is, a topological image of a polygonal line) such that there are circles outside of this arc that cannot possibly be contracted to a point without meeting it? There are also closed surfaces of genus zero which possess an analogous property. In other words, one can construct a topological image of a sphere and an ordinary circle in its interior in such a way that the circle may not be contracted to a point wholly inside the surface.[4]

---

[4] The common boundaries of three or more regions were discovered by Brouwer. We sketch here his construction for the case of three regions; the general case

FIG. 4

proceeds completely analogously. Imagine an island in the sea and on it a cold and a warm lake. The following work program is to be carried out on the island. In the

4. All of these phenomena were wholly unsuspected at the beginning of the current century; the development of set-theoretic methods in topology first led to their discovery and, consequently, *to a substantial extension of our idea of space.* However, let me at once issue the emphatic warning that the most important problems of set-theoretic topology are in no way confined to the exhibition of, so to speak, "pathological" geometrical structures; on the contrary they are concerned with something quite positive. I would formulate *the* basic problem of set-theoretic topology as follows:

To determine which set-theoretic structures have a connection with the intuitively given material of elementary polyhedral topology and hence deserve to be considered as geometrical figures—even if very general ones.

Obviously implicit in the formulation of this question is the problem of a systematic investigation of structures of the required type, particularly with reference to those of their properties which actually enable us to recognize the above mentioned connection and so bring about the geometrization of the most general set-theoretic-topological concepts.

---

course of the first hour canals are to be dug, one from the sea, one from the warm lake, and one from the cold lake, in such a way that neither salt and fresh nor warm and cold water come into contact with one another, and so that at the end of the hour every point of land is at a distance of less than one kilometer from each kind of water (i.e. salt, cold and warm). In the next half hour, each of the canals is to be continued so that the different kinds of water remain separated, and at the end of the work the distance of every point from each kind of water is less than one-half kilometer. In an analogous manner the work for the next 1/4, 1/8, 1/16, ..., hour is continued. At the end of the second hour, the dry land forms a closed set F nowhere dense in the plane, and arbitrarily near to each of its points there exists sea water as well as cold and warm fresh water. The set F is the common boundary of three regions: the sea, the cold lake and the warm lake (extended by their corresponding canals). [This example is due essentially to the Japanese mathematician Yoneyama, *Tohoku Math. Journal*, Vol. 12 (1917) p. 60.] (Fig. 4).

The singular curves and surfaces in R^3 which are also mentioned have been constructed by Antoine [*J. Math. pures appl.*, Vol. (8) 4 (1921), pp. 221-325.] Also Alexander: *Proc. Nat. Acad. U.S.A.* Vol. 10 (1924), pp. 6-12. Concerning the invariance of dimension, the tiling theorem and related questions see, in addition to Brouwer's classical work [*Math. Ann.* Vols. 70, 71, 72; *J. reine angew. Math.* Vol. 142 (1913), pp. 146-152; *Amsterd. Proc.*, Vol. 26 (1923), pp. 795-800]; Sperner, *Abh. Sem. Hamburg*, Vol. 6 (1928), pp. 265-272; Alexandroff, *Ann. of Math.*, Vol (2) 30 (1928), pp. 101-187, as well as "Dimensionstheorie" [*Math. Ann.*, Vol. 106 (1932), pp. 161-238].

Shortly a detailed work on topology by Professor H. Hopf and the author will be published in which all branches of topology will be taken into account. [*Topologie*, published by J. Springer, 1935—A.E.F.].

The program of investigation for set-theoretic topology thus formulated is to be considered—at least in basic outline—as completely capable of being carried out: it has turned out that the most important parts of set-theoretic topology are amenable to the methods which have been developed in polyhedral topology.[5] Thus it is justified if in what follows we devote ourselves primarily to the topology of polyhedra.

I. Polyhedra, Manifolds, Topological Spaces.

5. We begin with the concept of a *simplex*. A zero-dimensional simplex is a point; a one-dimensional simplex is a straight line segment. A two-dimensional simplex is a triangle [including the plane region which it bounds—A.E.F.], a three-dimensional simplex is a tetrahedron. It is known and easy to show that if one considers all possible distributions of (non-negative) mass at the four vertices of a tetrahedron the point set consisting of the centers of mass of these distributions is precisely the tetrahedron itself; this definition extends easily to arbitrary dimension. We assume here that the $r + 1$ vertices of an r-dimensional simplex are not contained in an $(r - 1)$-dimensional hyperplane (of the R^n we are considering). One could also define a simplex as the smallest closed convex set which contains the given vertices.

Any $s + 1$ of the $r + 1$ vertices of an r-dimensional simplex ($0 \leq s \leq r$) define an s-dimensional simplex—an *s-dimensional face* of the given simplex (the zero-dimensional faces are the vertices). Then we mean by an *r-dimensional polyhedron*, a point-set of R^n which can be decomposed into r-dimensional simplexes in such a way that two simplexes of this decomposition either have no points in common or have a common face (of arbitrary dimension) as their intersection. The system of all of the simplexes (and their faces) which belong to a simplicial decomposition of a polyhedron is called a *geometrical complex*.

The dimension of the polyhedron is not only independent of the choice of the simplicial decomposition, but indeed it expresses a *topological*

---

[5] We defer these questions until sections **34** and **41**. Concerning the general standpoint appearing here and its execution, see the works of the author mentioned in the preceding footnote. The basic work on general point set theory and at the same time the best introduction to set-theoretic topology is *Mengenlehre* by Hausdorff. See also Menger, *Dimensionstheorie*.

invariant of the polyhedron; that is to say, two polyhedra have the same dimension if they are *homeomorphic* (if they can be mapped onto one another in a one-to-one and bicontinuous fashion).[6]

With the general viewpoint of topology in mind (according to which two figures—that is, two point sets—are considered equivalent if they can be mapped onto one another topologically), we shall understand a general or *curved* polyhedron to be any point set which is homeomorphic to a polyhedron (defined in the above sense, that is, composed of ordinary "rectilinear" simplexes). Clearly, curved polyhedra admit decomposition into "curved" simplexes (that is, topological images of ordinary simplexes); the system of elements of such a decomposition is again called a geometrical complex.

6. The most important of all polyhedra, indeed, even the most important structures of the whole of general topology, are the so-called *closed n-dimensional manifolds* M^n. They are characterized by the following two properties. First, the polyhedron must be connected (that is, it must not be composed of several disjoint sub-polyhedra); second, it must be "homogeneously *n*-dimensional" in the sense that every point p of M^n possesses a neighborhood[7] which can be mapped onto the *n*-dimensional cube in a one-to-one and bicontinuous fashion, such that the point p under this mapping corresponds to the center of the cube.[8]

7. In order to recognize the importance of the concept of manifold, it suffices to remark that most geometrical forms whose points may be

---

[6] One-to-one and bicontinuous mappings are called *topological mappings* or *homeomorphisms*. Properties of point sets which are preserved in such mappings are called *topological invariants*. The theorem just mentioned is another form of the Brouwer theorem on the invariance of dimension. (It will be proved in sections **29-32**.)

[7] The general concept of "neighborhood" will be further explained in section **8**. The reader who wishes to avoid this concept may take "neighborhood of a point of a polyhedron" to mean the set union of all simplexes of an arbitrary simplicial decomposition of a polyhedron which contain the given point in their interior or on their boundary.

[8] On the subject of manifolds see, chiefly, Veblen, *Analysis Situs*, second edition, 1931; Lefschetz, *Topology*, 1931 (both printed by the American Mathematical Society). Further references are Hopf, *Math. Ann.*, Vol. 100 (1928), pp. 579-608; Vol. 102 (1929), pp. 562-623; Lefschetz, *Trans. Amer. Math. Soc.*, Vol. 28 (1926), pp. 1-49; Hopf, *Jour. f. Math.*, Vol. 163 (1930), pp. 71-88. See also the literature given in footnote 41.

defined by n parameters are n-dimensional manifolds; to these structures belong, for example, phase-spaces of dynamical problems. These structures are, to be sure, only rarely defined directly as polyhedra; rather they appear—as the examples of phase-spaces or the structures of n-dimensional differential geometry already show—as *abstract spatial constructions*, in which a concept of continuity is defined in one way or another; it turns out here (and it can be proved rigorously under suitable hypotheses) that in the sense of the above mentioned definition of continuity the "abstract" manifold in question can be mapped topologically onto a polyhedron, and thus falls under our definition of manifold. In this way, the projective plane, which is first defined as an abstract two-dimensional manifold, can be mapped topologically onto a polyhedron of four-dimensional space without singularities or self-intersections.[9]

8. Just one step leads from our last remarks to one of the most important and at the same time most general concepts of the whole of modern topology—the concept of *topological space*. A topological space is nothing other than a set of arbitrary elements (called "points" of the space) in which a concept of continuity is defined. Now this concept of continuity is based on the existence of relations, which may be defined as local or neighborhood relations—it is precisely these relations which are preserved in a continuous mapping of one figure onto another. Therefore, in more precise wording, a topological space is a set in which certain subsets are defined and are associated to the points of the space as their neighborhoods. Depending upon which axioms these neighborhoods satisfy, one distinguishes between different types of topological spaces. The most important

---

[9] Perhaps the easiest way to embed the projective plane topologically in R^4 is as follows: first, it is easy to convince oneself that a conic section decomposes the projective plane into an elementary piece of surface (a homeomorph of a circular disc), and a region homeomorphic to a Möbius band: indeed, the part interior to the conic is the elementary piece of surface, while the exterior part is topologically equivalent to the Möbius band (this is seen most easily if—choosing the line at infinity—one thinks of the conic section as a hyperbola).

Then one considers four-dimensional space R^4 and in it an R^3. In the R^3 construct a Möbius band. If one now chooses a point 0 in R^4, outside the R^3, and connects it by line segments to all points on the boundary curve of the Möbius band, an elementary piece of surface is formed which is joined to the Möbius band along its boundary and together with it forms a surface which is homeomorphic to the projective plane.

among them are the so-called Hausdorff spaces (in which the neighborhoods satisfy the four well-known Hausdorff axioms).

These axioms are the following:

(a) To each point x there corresponds at least one neighborhood $U(x)$; each neighborhood $U(x)$ contains the point x.

(b) If $U(x)$ and $V(x)$ are two neighborhoods of the same point x, then there exists a neighborhood $W(x)$, which is a subset of both.

(c) If the point y lies in $U(x)$, there exists a neighborhood $U(y)$, which is a subset of $U(x)$.

(d) For two distinct points x, y there exist two neighborhoods $U(x)$, $U(y)$ without common points.

Using the notion of neighborhood, the concept of continuity can be immediately introduced: A mapping f of a topological space R onto a (proper or improper) subset of a topological space Y is called *continuous* at the point x, if for every neighborhood $U(y)$ of the point $y = f(x)$ one can find a neighborhood $U(x)$ of x such that all points of $U(x)$ are mapped into points of $U(y)$ by means of f. If f is continuous at every point of R, it is called continuous in R.

9. The concept of topological space is only one link in the chain of abstract space constructions which forms an indispensable part of all modern geometric thought. All of these constructions are based on a common conception of *space* which amounts to considering one or more systems of objects—points, lines, etc.—together with systems of axioms describing the relations between these objects. Moreover, this idea of a space depends only on these relations and not on the nature of the respective objects. Perhaps this general standpoint found its most fruitful formulation in Hilbert's *Grundlagen der Geometrie*; however, I would especially emphasize that it is by no means only for investigations of the foundations that this concept is of decisive importance, but for all directions of present-day geometry—the modern construction of projective geometry as well as the concept of a many-dimensional Riemannian manifold (and indeed, earlier still, the Gaussian intrinsic differential geometry of surfaces) may suffice as examples!

10. With the help of the concept of topological space we have at last found an adequate formulation for the general definition of a manifold:

A topological space is called a closed n-dimensional manifold if it is homeomorphic to a connected polyhedron, and furthermore, if its points possess neighborhoods which are homeomorphic to the interior of the n-dimensional sphere.

11. We will now give some examples of closed manifolds.

The only closed one-dimensional manifold is the circle.

The "uniqueness" is of course understood here in the topological sense: every one-dimensional closed manifold is homeomorphic to the circle.

The closed two-dimensional manifolds are the orientable (or two-sided) and non-orientable (or one-sided) surfaces. The problem of enumerating their topological types is completely solved.[10]

As examples of higher-dimensional manifolds—in addition to n-dimensional spherical or projective space—the following may be mentioned:

1. The three-dimensional manifold of line elements lying on a closed surface F. (It can be easily proved that if the surface F is a sphere then the corresponding M^3 is projective space.)

2. The four-dimensional manifold of lines of the three-dimensional projective space.

3. The three-dimensional *torus-manifold*: it arises if one identifies the diametrically opposite sides of a cube pairwise. The reader may confirm without difficulty that the same manifold may also be generated if one considers the space between two coaxial torus surfaces (of which one is inside the other) and identifies their corresponding points.

The last example is also an example of the so-called topological product construction—a method by which infinitely many different manifolds can be generated, and which is, moreover, of great theoretical importance. The product construction is a direct generalization of the familiar concept of coordinates. One constructs the product manifold $M^{p+q} = M^p \times M^q$ from the two manifolds M^p and M^q as follows: the points of M^{p+q} are the pairs $z = (x, y)$, where x is an arbitrary point of M^p and y an arbitrary point of M^q. A neighborhood $U(z_0)$ of the point $z_0 = (x_0, y_0)$ is defined to be the collection of all points $z = (x, y)$ such that x belongs to an arbitrarily

---

[10] See for example Hilbert and Cohn-Vossen, Sec. **48**, as well as Kerekjarto, *Topologie*, Chapter 5.

chosen neighborhood of x_0 and y belongs to an arbitrarily chosen neighborhood of y_0. It is natural to consider the two points x and y (of M^p and M^q respectively) as the two "coordinates" of the point (x, y) of M^{p+q}.

Obviously this definition can be generalized without difficulty to the case of the product of three or more manifolds. We can now say that the Euclidean plane [Not a closed manifold—A.E.F.] is the product of two straight lines, the torus the product of two circles, and the three-dimensional torus-manifold, the product of a torus surface with the circle (or the product of three circles). As further examples of manifolds one has, for example, the product $S^2 \times S^1$ of the surface of a sphere with the circle, or the product of two projective planes, and so on. The particular manifold $S^2 \times S^1$ may also be obtained if one considers the spherical shell lying between two concentric spherical surfaces S^2 and s^2 and identifies the corresponding points (i.e., those lying on the same radius) of S^2 and s^2. Only slightly more difficult is the proof of the fact that, if one takes two congruent solid tori and (in accordance with the congruence mentioned) identifies the corresponding points of their surfaces with one another, one likewise obtains the manifold $S^2 \times S^1$. Finally, one gets the product of the projective plane with the circle if in a solid torus one identifies each pair of diametrically opposite points of every meridian circle.

These few examples will suffice. Let it be remarked here that, at present, in contrast to the two-dimensional case, the problem of enumerating the topological types of manifolds of three or more dimensions is in an apparently hopeless state. We are not only far removed from the solution, but even from the first step toward a solution, a plausible conjecture.

II. *Algebraic Complexes.*

12. There is something artificial about considering a manifold as a polyhedron: the general idea of the manifold as a homogeneous structure of n-fold extent, an idea which goes back to Riemann, has nothing intrinsically to do with the simplicial decompositions which were used to introduce polyhedra. Poincaré, who undertook the first systematic topological study of manifolds, and thus changed topology from a collection of mathematical curiosities into an independent and significant branch of geometry, originally defined manifolds analytically with the aid of systems of equations. However, within only four years after the appearance of his first pioneering work[11] he took the point of view which today is known as *combinatorial* topology, and essentially amounts to the consideration of manifolds as polyhedra.[12] The advantage of this viewpoint lies in the fact that with its help the difficult—partially purely geometric, partially set-theoretic—considerations to which the study of manifolds leads are replaced by the investigation of a finite combinatorial model—namely, the system of the simplexes of a simplicial decomposition of the polyhedron (i.e., the geometrical complex)—which opens the way to the application of algebraic methods.

Thus, it turns out that the definition of a manifold which we use here is currently the most convenient, although it represents nothing more than a deliberate compromise between the set-theoretic concept of topological space and the methods of combinatorial topology, a *compromise* which, at present, can scarcely be called an *organic blending* of these two directions. The most important of the difficult problems[13] connected with the concept of manifold are by no means solved by the definition which we have adopted.

13.   We shall now turn to the previously mentioned algebraic methods of the topology of manifolds (and general polyhedra). The basic concepts in algebraic topology are those of *oriented simplex, algebraic complex* and *boundary* of an algebraic complex.

An oriented one-dimensional simplex is a *directed straight line segment* (a_0a_1), that is, a line which is traversed from the vertex a_0 to the vertex a_1. One can also say: an oriented one-dimensional simplex is one with a particular ordering of its endpoints. If we denote the oriented line (a_0a_1) by x^1 (where the superscript 1 gives the dimension), the oppositely oriented

---

[11] Analysis Situs, [*J. Ec. Polyt.* Vol. (2) 1 (1895) pp. 1-123].

[12] In the work: Complément à l'Analysis Situs, [*Palermo Rend.*, Vol. 13 (1899), pp. 285-343]. This work is to be considered as the first systematic presentation of combinatorial topology.

[13] These questions arise from the problem of the set-theoretic and the *combinatorial* characterizations of manifolds. The first problem is to establish necessary and sufficient set-theoretic conditions under which a topological space is homeomorphic to a polyhedron, that is, to establish necessary and sufficient conditions under which its points possess neighborhoods homeomorphic to R^n. The second seeks a characterization of those complexes which appear as simplicial decompositions of polyhedra and which possess the manifold property, (in other words, each of whose points possesses a neighborhood homeomorphic to R^n). Both problems remain unsolved and, doubtless, are among the most difficult questions in topology.

simplex (a_1a_0) will be denoted by $- x^1$. This same line considered without orientation we denote by

$$| x^1 | = | a_0a_1 | = | a_1a_0 | .$$

An oriented two-dimensional simplex—an oriented triangle—is a triangle with a particular sense of rotation or with a particular ordering of its vertices; at the same time, no distinction is made between orderings which differ from one another by an even permutation, so that $(a_0a_1a_2)$, $a_1a_2a_0)$, and $(a_2a_0a_1)$ represent one orientation, and $(a_0a_2a_1)$, $(a_1a_0a_2)$, and $(a_2a_1a_0)$ represent the other orientation of the triangle whose vertices are a_0, a_1, a_2. If one orientation of the triangle is denoted by $x^2 = (a_0a_1a_2)$, the other is called $- x^2$. The triangle considered without orientation will again be denoted by $| x^2 |$. The essential thing here is that in an oriented triangle the boundary is also to be understood as an oriented (directed) polygon. The boundary of an oriented triangle $(a_0a_1a_2)$ is the collection of oriented lines (a_0a_1), (a_1a_2), (a_2a_0). If one denotes the boundary of x^2 by $\dot{x}^2$, then our last statement is expressed by the formula

(1') $$\dot{x}^2 = (a_0a_1) + (a_1a_2) + (a_2a_0),$$

or equivalently[14]

(1) $$\dot{x}^2 = (a_1a_2) - (a_0a_2) + (a_0a_1).$$

We may also say that in the boundary of x^2, the sides (a_1a_2) and (a_0a_1) appear with the coefficient $+ 1$, and the side (a_0a_2), with the coefficient $- 1$.

14. Consider now any decomposition into triangles (or *triangulation*) of a two-dimensional polyhedron P^2. The system comprised of the triangles together with their edges and vertices forms what we called in Sec. **5** a two-dimensional geometrical complex K^2. Now, we choose a particular (but completely arbitrary) orientation x_i^2 of any one of the triangles $| x_i^2 |$, $1 \leq i \leq \alpha_2$[15], of our complex; in a similar way we choose *any*

---

[14] If one imagines $x^2 = (a_0a_1a_2)$ as a symbolic product of three "variables", a_0, a_1, a_2, one may write

$$\dot{x}^2 = \sum_{i=0}^{2} (- 1)^i \frac{\partial x^2}{\partial a_i} .$$

[15] Where α_2, α_1, α_0 denote the number of two-, one-, or zero-dimensional elements of the geometrical complex.

particular orientation x_j^1 of one of the sides $|\, x_j^1 \,|$, $1 \leq j \leq \alpha_1$[15]. The system of all x_i^2 we call an *oriented two-dimensional complex* C^2, that is, an orientation of the geometrical complex K^2. For the oriented complex C^2 we use the notation

$$C^2 = \sum_{i=1}^{\alpha_2} x_i^2.$$

In order to indicate that C^2 is the result of orienting the complex K^2, we shall sometimes write $|\, C^2 \,| = K^2$.

The boundary of each oriented triangle x_i^2 can now be represented by a linear form

$$(2) \qquad \dot{x}_i^2 = \sum_{j=1}^{\alpha_1} t_i^j x_j^1,$$

where $t^j = +1$, -1 or 0, according to whether the oriented line x_j^1 occurs in the boundary of the oriented triangle x_i^2 with the coefficient $+1$, -1, or not at all.

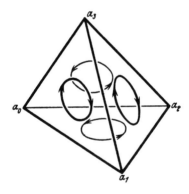

Fig. 5

If one sums equation (2) over all i, $1 \leq i \leq \alpha_2$, one obtains

$$\sum_{i=1}^{\alpha_2} \dot{x}_i^2 = \sum_{i=1}^{\alpha_2} \sum_{j=1}^{\alpha_1} t_i^j x_j^1 = \sum_{j=1}^{\alpha_1} u^j x_j^1, \qquad u^j = \sum_{i=1}^{\alpha_2} t_i^j.$$

The above expression

$$\sum_{j=1}^{\alpha_1} u^j x_j^1$$

is called the *boundary of the oriented complex* C^2 and is denoted by $\dot{C}^2$.

Examples. 1. Let K be the system composed of the four triangular faces of a tetrahedron; let the orientation of each of the faces be as indicated by the directions of the arrows in Fig. 5.

The boundary of the resulting oriented complex

$$C^2 = x_1^2 + x_2^2 + x_3^2 + x_4^2$$

equals zero, because each edge of the tetrahedron appears in the two triangles of which it is a side with different signs. In formulas:

$$x_1^2 = (a_0 a_1 a_2), \qquad x_2^2 = (a_1 a_0 a_3), \qquad x_3^2 = (a_1 a_3 a_2), \qquad x_4^2 = (a_0 a_2 a_3),$$

and

$$x_1^1 = (a_0 a_1), \qquad x_2^1 = (a_0 a_2), \qquad x_3^1 = (a_0 a_3), \qquad x_4^1 = (a_1 a_2),$$

$$x_5^1 = (a_1 a_3), \qquad x_6^1 = (a_2 a_3);$$

thus

$$\dot{x}_1^2 = + x_1^1 - x_2^1 \qquad + x_4^1$$

$$\dot{x}_2^2 = - x_1^1 \qquad + x_3^1 \qquad - x_5^1$$

$$\dot{x}_3^2 = \qquad\qquad - x_4^1 + x_5^1 - x_6^1$$

$$\dot{x}_4^2 = \qquad + x_2^1 - x_3^1 \qquad\qquad + x_6^1$$

$$\dot{C}^2 = \sum_{i=1}^{4} \dot{x}_i^2 = 0.$$

2. If one orients the ten triangles of the triangulation of the projective plane shown in Fig. 6 as indicated by the arrows, and puts

$$C^2 = \sum_{i=1}^{10} x_i^2,$$

then

(3) $$\dot{C}^2 = 2x_1^1 + 2x_2^1 + 2x_3^1.$$

The boundary of the oriented complex consists, therefore, of the projective line AA' (composed of the three segments x_1^1, x_2^1, x_3^1) *counted twice.* With another choice of orientations x_1^2, x_2^2, x_3^2, ..., x_{10}^2 of the ten triangles of this triangulation one would obtain another oriented complex

$$\sum_{i=1}^{10} x_i^2,$$

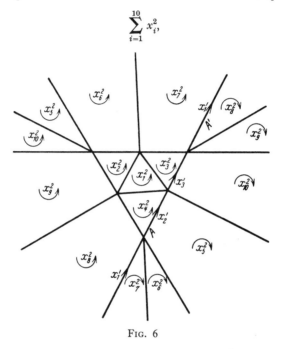

FIG. 6

and its boundary would be different from (3). Hence, it is meaningless to speak of the "boundary of the projective plane"; one must speak only of the boundaries of the various oriented complexes arising from different triangulations of the projective plane.

One can easily prove that no matter how one orients the ten triangles of Fig. 6, the boundary of the resulting complex

$$C^2 = \sum_{i=1}^{10} x_i^2$$

is never zero. In fact, the following general result (which can be taken as the definition of orientability of a closed surface) holds:

A closed surface is *orientable* if and only if one can orient the triangles of any of its triangulations in such a way that the oriented complex thus arising has boundary zero.

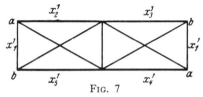

FIG. 7

3. In the triangulation and orientation for the Möbius band given in Fig. 7, we have:

$$\dot{C}^2 = 2x_1^1 + x_2^1 + x_3^1 + x_4^1 + x_5^1.$$

15. Oriented complexes and their boundaries serve also as examples of so-called *algebraic complexes*. A (two-dimensional) oriented complex, that is, a system of oriented simplexes taken from a simplicial decomposition of a polyhedron, was written by us as a linear form, $\Sigma\, x_i^2$; furthermore, as the boundary of the oriented complex $C^2 = \Sigma\, x_i^2$, there appeared a linear form $\Sigma\, u^j x_j^1$ whose coefficients are, in general, taken as *arbitrary* integers. Such linear forms are called *algebraic complexes*. The same considerations hold in the *n*-dimensional case if we make the general definition:

Definition I. An *oriented r-dimensional simplex* x^r is an *r*-dimensional simplex with an arbitrarily chosen ordering of its vertices,

$$x^r = (a_0 a_1 \ldots a_r),$$

where orderings which arise from one another by even permutations of the vertices determine the same orientation (the same oriented simplex), so that each simplex $|\, x^r\,|$ possesses two orientations, x^r and $-\, x^r$.[16]

Remark. Let x^r be an oriented simplex. Through the $r + 1$ vertices of x^r passes a unique *r*-dimensional hyperplane R^r (the R^r in which x^r lies), and to each *r*-dimensional simplex $|\, y^r\,|$ of R^r there exists a unique

---

[16] A zero-dimensional simplex has only one orientation, and thus it makes no sense to distinguish between x^0 and $|\, x^0\,|$.

orientation y^r such that one can map R^r onto itself by an affine mapping with a positive determinant in such a way that under this mapping the oriented simplex x^r goes over into the simplex y^r. Thus the orientation x^r of $| x^r |$ *induces* a completely determined orientation y^r for each simplex $| y^r |$ which lies in the hyperplane R^r containing x^r. Under these circumstances, one says that the simplexes x^r and y^r are *equivalently*—or *consistently—oriented* simplexes of R^r. One says also that *the whole coordinate space R^r is oriented by x^r*, which means precisely that from the oriented simplex x^r all r-dimensional simplexes of R^r acquire a fixed orientation. In particular, one can orient each r-dimensional simplex lying in R^r so that it has an orientation *equivalent* to that of x^r.

Definition II.   A linear form with integral coefficients $C^r = \Sigma\ t^i x_i^r$ whose indeterminants x_i are oriented r-dimensional *simplexes*, is called an *r-dimensional algebraic complex.*[17]

Expressed otherwise: an algebraic complex is a system of oriented simplexes, each of which is to be counted with a certain multiplicity (i.e., each one is provided with an integral coefficient). Here it will generally be assumed only that these simplexes lie in one and the same coordinate space R^n; we do not assume, however, that they are all taken from a definite simplicial decomposition of a polyhedron (i.e., a geometrical complex). On the contrary, the simplexes of an algebraic complex may, in general, intersect one another arbitrarily. In case the simplexes of an algebraic complex C^r belong to a geometrical complex (i.e., are obtained by orienting certain elements of a simplicial decomposition of a polyhedron), C^r is called an *algebraic subcomplex of the geometrical complex* (of the given simplicial decomposition) *in question*; here self-intersections of simplexes, of course, cannot occur. This case is to be considered as the most important.

16.   Algebraic complexes are to be considered as a higher-dimensional generalization of ordinary directed polygonal paths; here, however, the concept of polygonal path is taken from the outset in the most general sense: the individual lines may intersect themselves, and there may also exist lines which are traversed many times; moreover, one should not

---

[17] This definition also has meaning in the case $r = 0$. A zero-dimensional algebraic complex is a finite system of points with which some particular (positive, negative, or vanishing) integers are associated as coefficients. [In modern terminology, C^r would be called an (integral) r-dimensional chain—A.E.F.]

forget that the whole thing is to be considered algebraically, and a line which is traversed twice in opposite directions no longer counts at all. Furthermore, polygonal paths may consist of several pieces (thus no requirement of connectedness!). Thus, the two figures 8 and 9 represent polygonal paths which, considered as algebraic complexes, have the same structure (i.e., represent the same linear form).

Since the r-dimensional algebraic complexes of R^n may, as linear forms, be added and subtracted according to the usual rules of calculating with such symbols, they form an Abelian group $L_r(R^n)$. One can also consider, instead of the whole of R^n, a subspace G of R^n, for example; the r-dimensional algebraic complexes lying in it then form the group $L_r(G)$—a subgroup of $L_r(R^n)$.

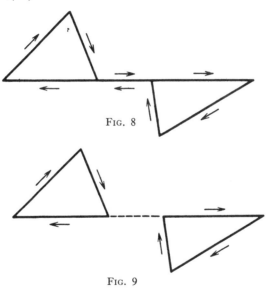

Fig. 8

Fig. 9

Also, the r-dimensional algebraic subcomplexes of a geometric complex K form a group—the group $L_r(K)$; it is the starting point for almost all further considerations. Before we continue with these considerations however, I would like to direct the attention of the reader to the fact that the concepts "polyhedron," "geometrical complex," and "algebraic complex" belong to entirely different logical categories: a polyhedron is a point set, thus a set whose elements are ordinary points of R^n; a geometrical complex is a (finite) set whose elements are simplexes, and, indeed, simplexes in the naive geometrical sense, that is, without orientation. An algebraic complex is not a set at all: it would be false to say that an algebraic complex is a set of oriented simplexes, since the essential thing about an algebraic complex is that the simplexes which appear in it are provided with coefficients and, therefore, in general, are to be counted with a certain multiplicity. This distinction between the three concepts, which often appear side by side, reflects the essential difference between the set-theoretic and the algebraic approaches to topology.

17. The boundary $\dot{C}^r$ of the algebraic complex $C^r = \Sigma\ t^i x_i^r$ is defined to be the algebraic sum of the boundaries of the oriented simplexes x_i^r, i.e., $\Sigma\ t^i \dot{x}_i^r$, where the boundary of the oriented simplex $x^r = (a_0 a_1 \ldots a_r)$ is the $(r-1)$-dimensional algebraic complex[18]

$$(4) \qquad \dot{x}^r = \sum_{i=0}^{r} (-1)^i\ (a_0 \ldots \hat{a}_i \ldots a_r),$$

where $\hat{a}_i$ means that the vertex a_i is to be omitted. In case the boundary of C^r is zero (for instance, in the case of example 1 of Sec. 11) C^r is called a cycle.[19] Thus in the group $L_r(R^n)$, and analogously in $L_r(K)$ and $L_r(G)$, the subgroup of all r-dimensional cycles $Z_r(R^n)$, or, respectively, $Z_r(K)$ and $Z_r(G)$, is defined.

We can now say (see Sec. 14): a closed surface is orientable if and only if one can arrange, by a suitably chosen orientation of any simplicial (i.e., in this case, triangular) decomposition of the surface, that the oriented complex given by this orientation is a cycle. Without change, this definition holds for the case of a closed manifold of arbitrary dimension. Let us remark immediately: orientability, which we have just defined as a property of a definite simplicial decomposition of a manifold, actually expresses a

---

[18] A zero-dimensional simplex has boundary zero; the boundary of a one-dimensional oriented simplex, i.e., the directed line segment $(a_0 a_1)$, is found by formula (4) to be $a_1 - a_0$, one endpoint having coefficient $+ 1$, one having coefficient $- 1$.

In the symbolic notation of footnote 13, one can write formula (4) in the form:

$$\dot{x}^r = \sum_{i=0}^{r} (-1)^i\ \frac{\partial x^r}{\partial a_i}\ .$$

[19] In particular, every zero-dimensional algebraic complex is obviously a cycle.

property of the manifold itself, since it can be shown that if *one* simplicial decomposition of a manifold satisfies the condition of orientability, the same holds true for *every* simplicial decomposition of this manifold.

Remark. If x^n and y^n are two equivalently oriented simplexes of R^n which have the common face $\mid x^{n-1} \mid$, then the face x^{n-1} (with some orientation) appears in $\dot{x}^n$ and $\dot{y}^n$ with the same or different signs according to whether the simplexes $\mid x^n \mid$ and $\mid y^n \mid$ lie on the same side or on different sides of the hyperplane R^{n-1} containing $\mid x^{n-1} \mid$. The proof of this assertion we leave to the reader as an exercise.

18. As is easily verified, the boundary of a simplex is a cycle. But from this it follows that the boundary of an arbitrary algebraic complex is also a cycle. On the other hand, it is easy to show that for each cycle $Z^r, r > 0$, in R^n there is an algebraic complex lying in this R^n which is bounded by Z^r:[20] indeed, it suffices to choose a point O of the space different from

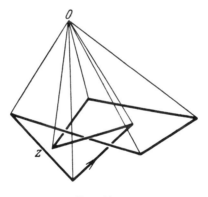

FIG. 10

all the vertices of the cycle Z^r and to consider the pyramid erected above the given cycle (with the apex at O) (Fig. 10). In other words, if

$$Z^r = \sum_{(i)} c^i x_i^r, \qquad \text{and} \qquad x_i^r = (a_0^i, a_1^i, ..., a_r^i),$$

---

[20] On the other hand, a zero-dimensional cycle in R^n bounds if and only if the sum of its coefficients equals zero (the proof is by induction on the number of sides of the bounded polygon).

then one defines the $(r + 1)$-dimensional oriented simplex x_i^{r+1} as

$$x_i^{r+1} = (O, a_0^i, ..., a_r^i)$$

and considers the algebraic complex

$$C^{r+1} = \sum_{(i)} c^i x_i^{r+1}.$$

The boundary of C^{r+1} is Z^r, since everything else cancels out.

If we consider, however, instead of the whole of R^n, some region G in R^n (or more generally, an arbitrary open set in R^n), then the situation is no longer so simple: a cycle of R^n lying in G need not bound in G. Indeed,

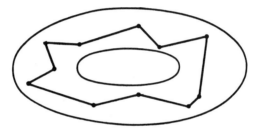

FIG. 11

if the region G is a plane annulus, then it is easy to convince oneself that there are cycles which do not bound in G (in this case closed polygons which encircle the center hole) (Fig. 11). Similarly, in a geometrical complex, there are generally some cycles which do not bound in the complex. For example, consider the geometrical complex of Fig. 12: the cycle ABC as well as the cycle abc obviously does not bound.

Consequently, one distinguishes the subgroups $B_r(G)$ of $Z_r(G)$, and $B_r(K)$ of $Z_r(K)$, of bounding cycles: the elements of $B_r(G)$, or $B_r(K)$, are cycles which bound some $[(r + 1)$-dimensional—A.E.F.] algebraic complex in G, or respectively, K.

In the example of the triangulation given in Fig. 6 of the projective plane, we see that it can happen that a cycle z does not bound in K, while a certain fixed integral multiple of it (i.e., a cycle of the form tz where t is an integer different from zero) does bound some algebraic subcomplex of K. We have, in fact, seen that the cycle $2x_1^1 + 2x_2^1 + 2x_3^1 = 2z^1$ (the "doubly-counted projective line") in the triangulation of Fig. 6 bounds, while in the same triangulation there is no algebraic complex which has

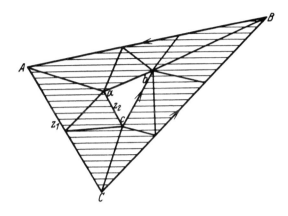

FIG. 12

the cycle $z^1 = (x_1 + x_2 + x_3)$ as its boundary. It is thus suitable to designate as *boundary divisors* all of those cycles z^r of K (of G) for which there exists a non-zero integer t such that tz bounds in K (in G). Since t may have the value 1, the true boundaries (i.e., bounding cycles) are included among the *boundary divisors*. The boundary divisors form, as is easily seen, a subgroup of the group $Z_r(K)$, or $Z_r(G)$, which we denote by $B'_r(K)$, or $B'_r(G)$; obviously, the group $B_r(K)$ is contained in the group $B'_r(K)$.

19. If z^r bounds in K (in G) we also say that z^r is *strongly homologous* to zero in K (in G), and we write $z^r \sim 0$ (in K or in G); if z^r is a boundary divisor of K (of G), we say that z is *weakly homologous* to zero and write $z \approx 0$ (in K or in G).

If two cycles of a geometrical complex K (or of region G) have the property that the cycle $z^r_1 - z^r_2$ is homologous to zero, one says that the cycles z^r_1 and z^r_2 are *homologous to one another*; this definition is valid for strong as well as for weak homology, so that one has the relations $z^r_1 \sim z^r_2$ and $z^r_1 \approx z^r_2$. Examples of these relations are given in Fig. 12 ($z_1 \sim z_2$) and in the following figures.

In Figs. 15 and 16, G is the region of three-dimensional space which is complementary to the closed Jordan curve S or, resp., to the lemniscate Λ.

20. Thus, the group $Z_r(K)$ falls into so-called *homology* classes, that is, into classes of cycles which are homologous to one another; there are in general both weak and strong homology classes, according to whether the concept of homology is meant to be weak or strong. If one again takes for K the geometrical complex of Fig. 6, then there are two strong homology classes of dimension one, for every one-dimensional cycle of K is either homologous to zero (that is, belongs to the zero-class) or homologous to the projective line (that is, say, the cycle $x_1 + x_2 + x_3$). Since every one-dimensional cycle of K in our case is a boundary divisor, there is only one weak homology class—the zero class.

As for the one-dimensional homology classes of the complexes in Figs. 12 and 13, they may be completely enumerated if one notices that in Fig. 12

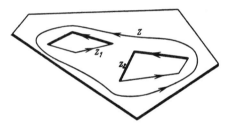

FIG. 13. $z \sim z_1 + z_2$ (in K).

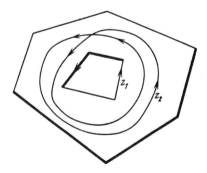

FIG. 14. $z_2 \sim 2z_1$ (in K).

every one-dimensional cycle satisfies a homology of the form $z \sim t z_1$, and, in Fig. 13, a homology of the form $z \sim u z_1 + v z_2$, where t, u, and v are

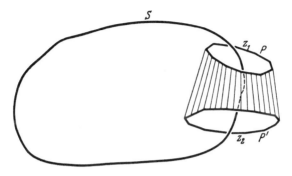

FIG. 15. $z_1 \sim z_2$ (in G).

integers; furthermore, the strong homology classes coincide with the weak in both complexes (for there are no boundary divisors which are not at the same time boundaries).

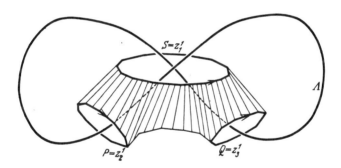

FIG. 16. $z_1 \sim z_2 + z_3$ (in G).

If ζ_1 and ζ_2 are two homology classes and z_1 and z_2 are arbitrarily chosen cycles in ζ_1 and ζ_2, respectively, then one denotes by $\zeta_1 + \zeta_2$ the homology class to which $z_1 + z_2$ belongs. This definition for the sum of two homology classes is valid because, as one may easily convince oneself, the homology class designated by $\zeta_1 + \zeta_2$ does not depend on the particular choice of the cycles z_1 and z_2 in ζ_1 and ζ_2.

The r-dimensional homology classes of K therefore form a group—the so-called factor group of $Z_r(K)$ modulo $B_r(K)$, or modulo $B'_r(K)$; it is called the *r-dimensional Betti group* of K. Moreover, one differentiates between the *full* and the *free* (or *reduced*) Betti groups—the first corresponds to the strong homology concept [it is, therefore, the factor group $Z_r(K)$ modulo $B_r(K)$, denoted $H_r(K)$], while the second is the group of the weak homology classes [the factor group $Z_r(K)$ modulo $B'_r(K)$, denoted $F_r(K)$].[21] For examples see Sec. **44**.

From the above discussion it follows that the full one-dimensional Betti group of the triangulation of the projective plane given in Fig. 6 is a finite group of order two; on the other hand, the free (one-dimensional) Betti group of the same complex is the zero group. The one-dimensional Betti group of the complex K (Fig. 12) is the infinite cyclic group, while in Fig. 13 the group of all linear forms $u\zeta_1 + v\zeta_2$ (with integral u and v) is the one-dimensional Betti group. In the latter two cases the full and reduced Betti groups coincide.

From simple group-theoretical theorems it follows that the full and the reduced Betti groups (of any given dimension r) have the same rank; that is, the maximal number of linearly independent elements which can be chosen from each group is the same. This common rank is called the *r*-dimensional *Betti number*[22] of the complex K. The one-dimensional Betti number for the projective plane is zero; for Figs. 12 and 13 it is, respectively, 1 and 2.

21. The same definitions are valid for arbitrary regions G contained in R^n. It is especially important to remember that, while in the case of a geometrical complex all of the groups considered had a finite number of generators, this is by no means necessarily the case for regions of R^n. Indeed, the region complementary to that consisting of infinitely many circles converging to a point (Fig. 18) has, as one may easily see, an infinite one-dimensional Betti number (consequently, the one-dimensional Betti

---

[21] [In fact, one may write $H_r(K) = F_r(K) \oplus T_r(K)$ where $T_r(K)$ is the subgroup of $H_r(K)$ consisting of the elements of finite order; the so-called torsion subgroup of $H_r(K)$. In current usage, the group $H_r(K)$ is more often referred to as the r-dimensional homology group rather than the r-dimensional Betti group.— A.E.F.]

[22] The reader will be able to prove easily that the zero-dimensional Betti number of a complex K equals the number of its *components* (i.e., the number of disjoint pieces of which the corresponding polyhedron is composed).

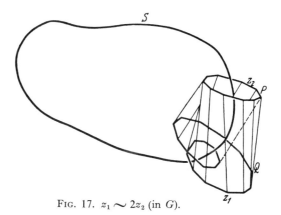

FIG. 17. $z_1 \sim 2z_2$ (in G).

group does not have finite rank, therefore, certainly not a finite number of generators).

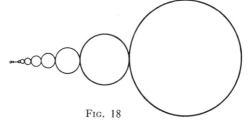

FIG. 18

22. The presentation of the basic concepts of the so-called *algebraic* topology[23] which we have just given is based on the concept of the oriented simplex. In many questions, however, one does not need to consider the orientation of the simplex at all—and can still use the algebraic methods extensively. In such cases, moreover, all considerations are much simpler, because the problem of sign (which often leads to rather tedious calculations) disappears. The elimination of orientation throughout, wherever it is actually possible, leads to the so-called "modulo 2" theory in which all coefficients of the linear forms that we have previously considered are replaced by their residue classes modulo 2. Thus, one puts the digit 0

---

[23] We prefer this expression to the otherwise customary term "combinatorial" topology, since we consider much broader applications of algebraic methods and concepts than the word "combinatorial" would include.

in place of any even number, the digit 1 in place of any odd number, and calculates with them in the following way:

$$0 + 0 = 0, \qquad 0 + 1 = 1 + 0 = 1, \qquad 1 + 1 = 0,$$
$$0 - 1 = 1 - 0 = 1,$$
$$0 - 0 = 1 - 1 = 0.$$

In particular, an algebraic complex mod 2 is a linear form whose indeterminants are simplexes considered without orientation and with coefficients 0 and 1.[24] The boundary of a simplex x^n appears in the theory mod 2 as a complex mod 2 which consists of all $(n - 1)$-dimensional faces of the simplex x^n. Hence, the boundary mod 2 of an arbitrary complex C^n is defined as the sum (always mod 2) of the boundaries of the individual n-dimensional simplexes of C^n. One can also say that the boundary mod 2 of C^n consists of those and only those $(n - 1)$-dimensional simplexes of C^n with which an odd number of n-dimensional simplexes are incident. The reader may easily construct examples which illustrate what has been said.

The concepts of cycle, homology, and Betti group mod 2 can be introduced exactly as in the "oriented" case. It should be especially noticed that all of our groups $L_r(K)$, $Z_r(K)$, $H_r(K)$ and so on, are now finite groups (which we shall denote by $L_r(K; Z_2)$, $Z_r(K; Z_2)$, $H_r(K; Z_2)$ etc. [where Z_2 is the group of residue classes mod 2—A.E.F.]), because we now have, throughout, linear forms in finitely many indeterminants whose coefficients take only the two values 0 and 1. The triangulation of the projective plane given in Fig. 6 can serve as an example of a two-dimensional cycle mod 2, for—considered as an algebraic complex mod 2—it obviously has vanishing boundary. In the case of an n-dimensional complex K^n [i.e. a complex consisting of simplexes of dimension $\leq n$—A.E.F.], just as $Z_n(K^n)$ is isomorphic to $H_n(K^n)$ [since there are no $(n + 1)$-dimensional simplexes, there can be no n-dimensional boundaries—A.E.F.], so is $Z_n(K^n; Z_2)$ isomorphic to $H_n(K^n; Z_2)$; therefore, the two-dimensional Betti group mod 2 for the projective plane is different from zero (its order is 2); the one-dimensional Betti group mod 2 in the case of the projective plane is also of order 2.

---

[24] One can consider geometrical complexes as a special case of the algebraic complexes modulo 2, if one interprets the coefficient 1 as signifying the occurrence, and the coefficient 0 as signifying the non-occurrence, of a simplex in a complex. This remark allows us to apply to geometrical complexes theorems which are proved for algebraic complexes.

Finally, one can also introduce the concept of the r-dimensional Betti number mod 2; this is the *rank mod 2* of the group $H_r(K; Z_2)$, that is, the greatest number of elements u_1, u_2, ..., u_s of this group such that a relation of the form

$$t_1 u_1 + t_2 u_2 + ... + t_s u_s = 0$$

is satisfied only if all t_i vanish (where the t_i take only the values 0 and 1).

The zero-, one-, and two-dimensional Betti numbers of the projective plane (mod 2) all have the value 1.[25]

23. We close our algebraic-combinatorial considerations with the concept of *subdivision*. If one decomposes each simplex of a geometrical complex K into ("smaller") simplexes such that the totality of all simplexes thus obtained again forms a geometrical complex K_1, then K_1 is called a *subdivision of K*. If K consists of a single simplex, then the elements of the subdivision which lie on the boundary of the simplex form a subdivision of the boundary. From this it follows that if K^n is a geometrical complex, K_1^n its subdivision, and K^r the complex consisting of all the r-dimensional elements of K^n [together with all of their faces—A.E.F.] ($r \leq n$), the totality of those elements of K_1^n which lie on simplexes of K^r forms a subdivision of K^r.

One can speak of subdivisions of algebraic complexes; we shall do this for the most important special case, in which the algebraic complex $C^n = \Sigma\, t^i x_i^n$ is an algebraic subcomplex of a geometrical complex. Then it is also true that the totality of all simplexes (considered without coefficients or orientation) of C^n forms a geometrical complex K^n. Let K_1^n be a

---

[25] The theory modulo 2 is due to Veblen and Alexander; it plays a very important role in modern topology, and has also prepared the way for the most general formulation of the concept of "algebraic complex": If J is any commutative ring with identity (see, for example, Van der Waerden: *Modern Algebra*, Chapter 3), we mean by an *algebraic complex of coefficient domain J* a linear form whose indeterminants are oriented simplexes whose coefficients are elements of the ring J. Then one defines boundaries, cycles, homology, etc., exactly as before but with respect to the ring J; in particular, the coefficient 1 (or -1) is now to be interpreted as an element of the ring (which, indeed, according to the hypothesis, contains an identity). If J is the ring of residue classes modulo m we speak of algebraic complexes modulo m. These complexes are gaining more and more significance in topology. Of greater importance as a coefficient domain is the set R of rational numbers; in particular, the cycles which we have called boundary divisors in Sec. **18** are nothing else but the cycles with integer coefficients which bound in K with respect to R (but not necessarily with respect to the ring of integers).

subdivision of K^n, and $|y^n|$ some simplex of C^n; then $|y^n|$ lies on some particular simplex $|x_i^n|$ of C^n. We now orient the simplex $|y^n|$ the same as x_i^n (see Sec. 15), and give it the coefficient t^i. In this way, we obtain an algebraic complex which is called a subdivision of the algebraic complex C^n. One can easily see that the boundary of the subdivision C_1^n of C^n is a subdivision of the boundary of C^n. (Considered modulo 2, the process gives nothing beyond the subdivision of a geometrical complex.)

III. *Simplicial Mappings and Invariance Theorems.*

24. If we review what has been said up till now, we see that the discussion has turned around two main concepts: *complex* on the one hand and *topological space* on the other. The two concepts correspond to the two interpretations of the concept basic to all of geometry—the concept of geometrical figure. According to the first interpretation, which has been inherent in synthetic geometry since the time of Euclid, a figure is a finite system of (generally) heterogeneous elements (such as points, lines, planes, etc., or simplexes of different dimensions) which are combined with one another according to definite rules—hence, a *configuration* in the most general sense of the word. According to the second interpretation, a figure is a *point set*, usually an *infinite* collection of homogeneous elements. Such a collection must be organized in one way or another to form a geometrical structure—a figure or space. This is accomplished, for example, by introducing a coordinate system, a concept of distance, or the idea of neighborhoods.[26]

As we mentioned before, in the work of Poincaré both interpretations appear simultaneously. With Poincaré the combinatorial scheme never becomes an end in itself; it is always a tool, an apparatus for the investigation of the "manifold itself," hence, ultimately a point set. Set-theoretic methods sufficed, however, in Poincaré's earliest work because his investigations touched only manifolds and slightly more general geometrical structures.[27] For this reason, and also in view of the great difficulties

---

[26] The set-theoretic interpretation of a figure also goes back to the oldest times— think, for example, of the concept of geometrical locus. This interpretation became prominent in modern geometry after the discovery of analytic geometry.

[27] Of course, in the fields of differential equations and celestial mechanics, the works of Poincaré already lead us very close to the modern formulation of questions in set-theoretic topology.

connected with the general formulation of the concept of manifold, one can hardly speak of an intermingling or merging of the two methods in Poincaré's time.

The further development of topology is marked at first by a sharp separation of set-theoretic and combinatorial methods: combinatorial topology had been at the point of believing in no geometrical reality other than the combinatorial scheme itself (and its consequences), while the set-theoretic direction was running into the same danger of complete isolation from the rest of mathematics by an accumulation of more and more specialized questions and complicated examples.

In the face of these extreme positions, the monumental structure of Brouwer's topology was erected which contained—at least in essence—the basis for the rapid fusion of the two basic topological methods which is presently taking place. In modern topological investigations there are hardly any questions of great importance which are not related to the work of Brouwer and for which a tool cannot be found—often readily applicable—in Brouwer's collection of topological methods and concepts.

In the twenty years since Brouwer's work, topology has gone through a period of stormy development, and we have been led—mainly through the great discoveries of the American topologists[28]—to the present "flowering" of topology, in which analysis situs—still far removed from any danger of being exhausted—lies before us as a great domain developing in close harmony with the most varied ideas and questions of mathematics.

At the center of Brouwer's work stand the *topological invariance theorems*. We collect under this name primarily theorems which maintain that if a certain property belonging to geometrical complexes holds for one simplicial decomposition of a polyhedron, then it holds for all simplicial decompositions of homeomorphic polyhedra. The classical example of such an invariance theorem is Brouwer's theorem on the invariance of dimension: *if an n-dimensional complex K^n appears as a simplicial decomposition of a polyhedron P, then every simplicial decomposition of P, as well as every simplicial decomposition of a polyhedron P_1 which is homeomorphic to P, is likewise an n-dimensional complex.*

Along with the theorem on invariance of dimension we mention as a

---

[28] Alexander, Lefschetz and Veblen in topology itself, and Birkhoff and his successors in the topological methods of analysis.

second example the *theorem on the invariance of the Betti groups* proved by Alexander: if K and K_1 are simplicial decompositions of two homeomorphic polyhedra P and P_1, then every Betti group of K is isomorphic to the corresponding Betti group of K_1.[29]

25. In the proof of the invariance theorems one uses an important new device—the *simplicial mappings* and *simplicial approximations* of continuous mappings introduced by Brouwer. Simplicial mappings are the higher-dimensional analogues of piecewise linear functions, while the simplicial approximations of a continuous mapping are analogous to linear interpolations of continuous functions. Before we give a precise formulation of these concepts, we remark that their significance extends far beyond the proof of topological invariance: namely, they form the basis of the whole general theory of continuous mappings of manifolds and are—together with the concepts of topological space and complex—among the most important concepts of topology.

26. To each vertex a of the geometrical complex K let there be associated a vertex $b = f(a)$ of the geometrical complex K' subject to the following restrictions: if a_1, ..., a_s are vertices of a simplex of K, then there exists in K' a simplex which has as its vertices precisely $f(a_1)$, ..., $f(a_s)$ (which, however, need not be distinct). From this condition it follows that to each simplex of K there corresponds an (equal- or lower-dimensional) simplex of K'.[30] One obtains in this way a mapping f of the complex K into the

---

[29] The scope of these two theorems is not lessened if one assumes that K and K_1 are two curved simplicial decompositions of one and the same polyhedron, for under a topological mapping an (arbitrary, also curved) simplicial decomposition of P_1 goes over into an (in general, curved) simplicial decomposition of P. One could, on the other hand, limit oneself to rectilinear simplicial decompositions of ordinary ("rectilinear") polyhedra, but then *both* polyhedra P and P_1 must be considered. If, indeed, P is an arbitrary polyhedron in a (curved) simplicial decomposition K_1, then there is a topological mapping of P into a sufficiently high dimensional Euclidean space in which P goes over into a rectilinear polyhedron P', and K, into its rectilinear simplicial decomposition K'.

[30] If one thinks of K as an algebraic complex modulo 2, then it is found to be convenient in the case where a simplex $| x^r |$ of K is mapped onto a lower-dimensional simplex of K' to say that *the image of $| x^r |$ is zero* (i.e., as an r-dimensional simplex, it vanishes).

complex K'. [31] A mapping of this kind from K to K' is called a *simplicial mapping of one geometrical complex into the other.*

27. If now, $x^r = (a_0 a_1 \ldots a_r)$ is an oriented simplex of K, then two cases are to be distinguished: either the image points $b_0 = f(a_0)$, ..., $b_r = f(a_r)$ are distinct vertices of K', in which case, we set $f(x^r) = (b_0 b_1 \ldots b_r)$; or else, at least two of the image points b_i, b_j coincide, in which case, we define $f(x^r) = 0$. *Thus the simplicial image of an oriented simplex will be either an oriented simplex of the same dimension or zero.* [32]

Now let an algebraic subcomplex $C^r = \Sigma\, t^i x_i^r$ of the complex K be given. According to what was just said, the simplicial mapping f of K into K' yields a well-defined image $f(x^r)$ for each oriented simplex x^r, where $f(x^r)$ is either an oriented r-dimensional simplex of K', or zero. Consequently, $f(C^r) = \Sigma\, t^i f(x_i^r)$ is a uniquely determined (perhaps vanishing) r-dimensional algebraic subcomplex of K' which is called *the image of C^r under the simplicial mapping of K into K'.* [33]

28. From these definitions follows easily the intuitive and extremely important

1. *Conservation Theorem. If the oriented simplex x^r is simplicially mapped into K', then $f(\dot{x}^r) = [f(x^r)]\dot{}$.*

From this by simple addition:

$$f(\dot{C}^r) = [f(C^r)]\dot{}.$$

In words: *the image of the boundary* (of an arbitrary algebraic complex) *is* (for every simplicial mapping) *equal to the boundary of the image.*

From the first conservation theorem follows without difficulty the extraordinarily important

---

[31] If to each element of the set M there corresponds an element of the set N, then one speaks of a mapping of the set M *into* the set N. The mapping is a mapping of M *onto* N if every element of N is the image of at least one element of M.

[32] The geometrical meaning of the occurrence of zero is clear: if two vertex images coincide, then the image simplex is degenerate; that is, it vanishes if one considers it as an r-dimensional simplex. The same mapping convention also holds when a non-oriented simplex is interpreted as an element of an algebraic complex modulo 2 (see footnote 30).

[33] One can also speak directly of the simplicial mapping f of the algebraic complex C^r into (the geometrical complex) K'.

2. *Conservation Theorem.* [34]   *If the algebraic complex C^n is simplicially mapped into the complex consisting of a single simplex $| x^n |$, and if $f(C^n) = \dot{x}^n$* (where x^n is some particular orientation of the simplex $| x^n |$), *then*

$$f(C^n) = x^n.$$

For, on the one hand, it is necessary that $f(C^n) = tx^n$ (where t is an integer which, a priori, could be zero), while, on the other hand, according to the assumption, $f(\dot{C}^n) = \dot{x}^n$, and by the first conservation theorem, $f(\dot{C}^n) = t\dot{x}^n$. Therefore, it must be that $t = 1$, q.e.d.

As an immediate application of the second conservation theorem we prove the following remarkable fact:

3. *Conservation Theorem.   Let C^n be an arbitrary* (algebraic) *complex, and C_1^n a subdivision of C^n. To each vertex a of C_1^n we let correspond a completely arbitrary vertex* f(a) *of that simplex of C^n which contains the point a in its interior;* [35] *then, for such a simplicial mapping* f *of the complex C_1^n it follows that*

$$f(C_1^n) = C^n.$$

Proof : For $n = 0$, the theorem is trivially true. We assume that it is proved for all $(n - 1)$-dimensional complexes, and consider an n-dimensional complex C^n. Let x_i^n be a simplex of $C^n = \Sigma\, t^i x_i^n$, and let X_i^n be the subdivision of x_i^n which is given by C_1^n. The mapping f of the boundary of X_i^n obviously fulfills the assumptions of our theorem, so that (because of the assumption of its validity for $n - 1$) $f(\dot{X}_i^n) = \dot{x}_i^n$; therefore, by the second conservation theorem, $f(X_i^n) = x_i^n$. Summing this over all simplexes x_i^n, one has

$$f(C_1^n) = f\Big(\sum_i t^i X_i^n\Big) = \sum_i t^i x_i^n = C^n, \text{ q.e.d.}$$

Remark.   Clearly, all three conservation theorems together with the proofs given here are valid mod 2; in this case, they may be considered as statements concerning geometrical complexes. [36] It is recommended that the reader verify this by examples—it suffices to choose a triangle for C^n, and an arbitrary subdivision of it for C_1^n.

---

[34] See Alexander, "Combinatorial Analysis Situs," *Trans. Amer. Math. Soc.,* Vol. 28 (1926), p. 328. Hopf, *Nachr. d. Ges. d. Wiss. Gttg.,* 1928, p. 134.

[35] In particular, if a is not only a vertex of C^n but also a vertex of C^n then our condition means that $f(a) = a$.

[36] They are valid quite generally for an arbitrary coefficient domain.

29. We apply the third conservation theorem to the proof of the tiling theorem, already mentioned in Sec. 1; however, we shall now formulate it not for a cube but for a simplex:

For sufficiently small $\epsilon > 0$, every ϵ-covering[37] of an n-dimensional simplex has order $\geq n + 1$.

First, we choose ϵ so small that there is no set with a diameter less than ϵ which has common points with all $(n - 1)$-dimensional faces of $|x^n|$. In particular, it follows that no set with a diameter less than ϵ can simultaneously contain a vertex a_i of $|x^n|$ and a point of the face $|x_i^{n-1}|$ opposite to the vertex a_i. Now let

$$(1) \qquad\qquad F_0, F_1, F_2, ..., F_s$$

be an ϵ-covering of $|x^n|$. We assume that the vertex a_i, $i = 0, 1, ..., n$ lies in F_i.[38] If there are more than $n + 1$ sets F_i, then we consider some set $F_j, j > n$, and proceed as follows: we look for a face $|x_i^{n-1}|$ of $|x^n|$ which is disjoint from F_j (such a face exists, as we have seen), strike out the set F_j from (1) and replace F_i by $F_i \cup F_j$, renaming this last set F_i. By this procedure, the number of sets in (1) is diminished by 1 without increasing the order of the system of sets in (1). At the same time, the condition that none of the sets F_i contains simultaneously a vertex and a point of the face opposite to the vertex will not be violated. By finite repetition of this process, we finally obtain a system of sets

$$(2) \qquad\qquad F_0, F_1, ..., F_n$$

containing the sequence of vertices, $a_0, a_1, ..., a_n$ of $|x^n|$, $a_i \varepsilon F_i$, with the property that no set contains both a vertex a_i and a point of $|x_i^{n-1}|$. Furthermore, the order of (2) is at most equal to the order of (1). It therefore suffices to prove that the order of (2) is equal to $n + 1$, i.e., to show that there is a point of $|x^n|$ which belongs to all sets of (2). As quite elementary convergence considerations show, the latter goal will be reached if we show that in each subdivision $|X^n|$ of $|x^n|$, no matter

---

[37] By an ϵ-*covering* of a closed set F one means a finite system $F_1, F_2, ..., F_s$ of closed subsets of F, which have as their union the set F and which are less than ϵ in diameter. The order of a covering (or more generally, of an arbitrary finite system of point sets) is the largest number k with the property that there are k sets of the system which have at least one common point.

[38] According to our assumption, two different vertices cannot belong to the same set F_i; a vertex a_i can, however, be contained in sets of our covering other than F_i.

how fine, there is necessarily a simplex $|y^n|$ which possesses points in common with all the sets F_0, F_1, ..., F_n.

Let b be an arbitrary vertex of the subdivision $|X^n|$. Now b belongs to at least one of the sets F_i; if it belongs to several, then we choose a particular one, for example, the one with the smallest subscript. Let this be F_i, then we define $f(b) = a_i$. In this way, we get a simplicial mapping f of $|X^n|$ into $|x^n|$ which, I assert, satisfies the conditions of the third conservation theorem.[39] Indeed, if b is interior to the face $|x^r|$ of $|x^n|$, then $f(b)$ must be a vertex of $|x^r|$; because otherwise, if the whole simplex $|x^r|$, and in particular the point b, were to lie on the face $|x^{n-1}|$ of $|x^n|$ opposite to $a_i = f(b)$, then the point b could not belong to F_i. Since, according to the third conservation theorem (understood modulo 2), $|x^n| = f(|X^n|)$, there must be among the simplexes of $|X^n|$ at least one which will be mapped by f onto $|x^n|$ (and not onto zero);[39] the vertices of this simplex must lie successively in F_0, F_1, ..., F_n, q.e.d.[40]

30.   If F is a closed set, then the smallest number r with the property that F possesses for each $\epsilon > 0$ an ϵ-covering of order $r + 1$, is called the *general* or *Brouwer dimension* of the set F. It will be denoted by dim F. If F' is a subset of F, then clearly, dim $F' \le$ dim F. It is easy to convince oneself that two homeomorphic sets F_1 and F_2 have the same Brouwer dimension.

In order to justify this definition of general dimension, however, one must prove that for an r-dimensional (in the elementary sense) polyhedron P, dim $P = r$; *one would, thereby, also prove Brouwer's theorem on the invariance of dimension.* Now, it follows at once from the tiling theorem that for an r-dimensional simplex, and consequently for every r-dimensional polyhedron P, that dim $P \ge r$. For the proof of the reverse inequality, we have only to construct, for each $\epsilon > 0$, an ϵ-covering of P of order $r + 1$. Such coverings are provided by the so-called *barycentric coverings* of the polyhedron.

---

[39] We consider $|x^n|$ as an *algebraic complex mod 2*, so that footnotes 30 and 32 are valid.

[40] The above proof of the tiling theorem is due in essence to Sperner; the arrangement given here was communicated to me by Herr Hopf. We have carried through the considerations modulo 2, since the theorem assumes no requirement of orientation. The same proof is also valid verbatim for the oriented theory (and, in fact, with respect to any domain of coefficients).

31. First, we shall introduce the *barycentric subdivisions* of an n-dimensional complex K^n. If $n = 1$, the barycentric subdivision of K^1 is obtained by inserting the midpoints of the one-dimensional simplexes of which K^1 consists [i.e., if the midpoint of each 1-simplex of K^1 is called the *barycenter* of that simplex, then the barycentric subdivision of K^1 is the complex consisting of all the vertices of K^1 and all barycenters of K^1 together with the line segments whose endpoints are these points—A.E.F.]. If $n = 2$, the barycentric subdivision consists in dividing each triangle of K^2 into six triangles by drawing its three medians (Fig. 19). Suppose that the barycentric subdivision is already defined for all r-dimensional complexes, then define it for an $(r + 1)$-dimensional complex K by barycentrically subdividing the complex K', consisting of all r-dimensional simplexes (and their faces), and projecting the resulting subdivision of the boundary of each $(r + 1)$-dimensional simplex of K from the center of gravity (barycenter) of this simplex. It is easy to see by induction that:

1. Each n-dimensional simplex is subdivided barycentrically into $(n + 1)!$ simplexes.

2. Among the $n + 1$ vertices of an n-dimensional simplex $| y^n |$ of the barycentric subdivision K_1 of K,

(0) one vertex is also a vertex of K (this vertex is called the "leading" vertex of $| y^n |$),

(1) one vertex is the center of mass of an edge $| x^1 |$ of K (which possesses the leading vertex of $| y^n |$ as a vertex),

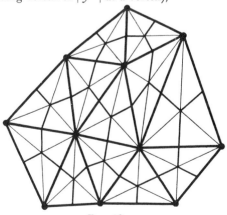

FIG. 19

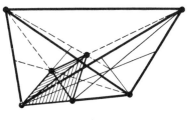

Fig. 20

(2) one vertex is the center of mass of a triangle $|\,x^2\,|$ of K (which is incident with the edge $|\,x^1\,|$),

. . . .

(n) one vertex is the center of mass of an n-dimensional simplex $|\,x^n\,|$ of K (which contains the previously constructed $|\,x^1\,|$, $|\,x^2\,|$, ..., $|\,x^{n-1}\,|$ among its faces). See Fig. 20.

32. One means by a *barycentric star* of K, the union of all simplexes of the barycentric subdivision K_1 of K which possess a fixed vertex a of K as their common (leading) vertex (see Fig. 19). The vertex a is called the *center* of the star.

It is easily shown that a point of a simplex $|\,x\,|$ of K can belong only to those barycentric stars whose centers are vertices of the simplex $|\,x\,|$. From this it follows that:

a) If certain barycentric stars B_1, B_2, ..., B_s have a common point p, then their centers are vertices of one and the same simplex of K (namely, that simplex which contains the point p in its interior).

b) There is a positive number $\epsilon = \epsilon(K)$ with the property that all points of the polyhedron P (whose simplicial decomposition is K) which are at a distance of less than ϵ from a simplex x of K can belong only to those barycentric stars which have their centers at the vertices of x. (This follows simply from the fact that all other stars are disjoint from x, and consequently have a positive distance from this simplex.)

The second of these two properties we will use later. We remark in passing that the converse of proposition a) is also true: if the centers of the barycentric stars B_1, B_2, ..., B_s lie at the vertices of one and the same simplex x of K, then they have a common point (namely, the center of mass of the simplex x). As a consequence, we have the following theorem:

Arbitrarily chosen barycentric stars of the complex K have a non-empty intersection if and only if their centers are vertices of a simplex of K.

In particular, the last statement includes the following corollary:

The system of all barycentric stars of an n-dimensional complex has order n + 1.

If one chooses a sufficiently fine simplicial decomposition K of an n-dimensional polyhedron P, then one can arrange it so that the barycentric stars of K are all of diameter less than ϵ, which therefore gives an ϵ-covering of P of order $n + 1$, q.e.d. The agreement of Brouwer's general dimension with the elementary geometrical dimension of a polyhedron, as well as the invariance of dimension are, hereby, completely proved.

33. *Remark I.* If a finite system of sets

(3) $$F_1,\ F_2,\ ...,\ F_s$$

and the system of vertices

$$a_1,\ a_2,\ ...,\ a_s$$

of a complex K are related to one another in such a way that the sets F_{i_0}, F_{i_1}, ..., F_{i_r} have a non-empty intersection if and only if the vertices a_{i_0}, a_{i_1}, ..., a_{i_r} belong to a simplex of K, then the complex K is called a *nerve* of the system of sets (3). Then one can formulate the theorem of the preceding section in the following way: *every complex K is a nerve of the system of its barycentric stars.*

34. *Remark II.* The previous remark leads us to the point at which the concept of complex attains its complete logical rigor and generality: it is exactly this example of the nerve of a system of sets which shows that the conceptual content of the word "complex" is, frequently, to a great extent independent of the "geometrical stuff" with which our concept operates. A complex, considered as the nerve of a system of sets (for example, the system of its own barycentric stars), is above all an abstract scheme which gives us information about the combinatorial structure of the system of sets. What the simplexes look like—whether they are "straight" or "curved"—what the nature of the vertices is, is completely immaterial to us; the only thing that concerns us is the structure of the network of vertices of the complex, that is, the manner in which the system of all vertices of the complex decomposes into the vertex-systems of the individual simplexes.[41]

Therefore, if one wants to define an *abstract* geometrical complex, it is most convenient to begin with a set E of (arbitrary) objects, which are called *vertices*; the set E we call a *vertex domain*. In E we then pick out certain finite subsets, which are called the *frames*; here the following two conditions must be satisfied:

1. Each individual vertex is a frame.

2. Every subset of a frame is a frame.

The number of vertices of a frame diminished by one will be called its *dimension*.

Finally, we suppose that to each frame is associated a new object—*the simplex spanned by the frame;* here we make no assumptions about the nature of this object; we are concerned only with the rule which associates to every frame a *unique simplex*. The dimension of the frame is called the *dimension of the simplex;* the simplexes spanned by the sub-frames of a given simplex x^n are called the *faces* of x^n. *A finite system of simplexes is called an abstract geometrical complex* of the given vertex domain.

Furthermore, one introduces the concept of orientation exactly as we have done previously. If this is done then the concept of an *abstract algebraic complex with respect to a definite domain of coefficients*[42] necessarily results.

From the fact that we formulate the concept of a complex abstractly, its range of application is substantially enlarged. As long as one adheres to the elementary geometrical conception of a complex as a simplicial decomposition of a polyhedron, one cannot free oneself from the impression that there is something arbitrary which is connected with the choice of this concept as *the* basic concept of topology: why should this particular notion, simplicial decomposition of polyhedra, constitute the central point of all topology? The abstract conception of a complex as a finite scheme

---

[41] This general standpoint was formulated with complete clarity for the first time in the works of the author: "Zur Begrundung der n-dimensionalen Topologie," *Math. Ann.* Vol. 94 (1925), pp. 296-308; "Simpliziale Approximationen in der allgemeinen Topologie," *Math. Ann.* Vol. 96 (1926), pp. 489-511; see also the correction in Vol. 101 (1929), pp. 452-456.

[42] *The general concept of algebraic complex thus arises by combining two different concepts: those of vertex and coefficient domains.* An algebraic complex is finally nothing but a prescription which associates to each simplex of a given vertex domain a definite element of the chosen coefficient domain.

which is, a priori, suitable for describing different processes (for example, the structure of a finite system of sets) helps to overcome this skepticism. Here, precisely those abstract complexes which are defined as nerves of finite systems of sets play a decisive role: that is, it can be shown that the topological investigation of *an arbitrary* closed set—therefore, the most general geometrical figure conceivable—can be *completely* reduced to the investigation of a sequence of complexes

(1) $$K_1^n, K_2^n, ..., K_h^n, ...$$

(n is the dimension of the set) related to one another by certain simplicial mappings. Expressed more exactly: for every closed set one can construct a sequence of complexes (1) and of simplicial mappings f_h of K_{h+1} into K_h ($h = 1, 2, ...$), (which also satisfies certain secondary conditions which, for the moment, need not be considered). Such a sequence of complexes and simplicial mappings is called a *projection spectrum*. Conversely, *every projection spectrum* defines in a certain way, which we cannot describe here, a uniquely determined class of mutually homeomorphic closed sets; moreover, one can formulate exact necessary and sufficient conditions under which two different projection spectra define homeomorphic sets. In other words: *the totality of all projection spectra falls into classes whose definition requires only the concepts "complex" and "simplicial mapping", and which correspond in a one-to-one way to the classes of mutually homeomorphic closed sets*. It turns out that the elements of a projection spectrum are none other than the nerves of increasingly finer coverings of the given closed sets. These nerves *can be considered as approximating complexes for the closed set*.[43]

35. We now go over to a brief survey of the proof of the invariance of the Betti numbers of a complex promised at the close of Sec. **25**. Since we are only going to emphasize the principal ideas of this proof, we shall forgo a proof of the fact that a geometrical complex[44] has the same Betti numbers as any one of its subdivisions.[45] We begin the proof with the following fundamental lemma:

---

[43] Concerning this, see P. Alexandroff, "Gestalt u. Lage abgeschlossener Mengen," *Ann. of Math.* Vol. 30 (1928), pp. 101-187.

[44] Until further notice, we are again dealing only with geometrical complexes, i.e., simplicial decompositions of (perhaps curved) polyhedra of a coordinate space.

[45] Concerning this, see for example Alexander, "Combinatorial Analysis Situs," *Trans. Amer. Math. Soc.*, Vol. 28 (1926), pp. 301-329.

Lebesgue's lemma. For every covering

(1) $$S = (F_1, \ F_2, \ ..., \ F_s)$$

of the closed set F, there is a number $\sigma = \sigma(S)$—*the Lebesgue number of the covering S*—with the following property: if there is a point a whose distance from certain members of the covering S—say $F_{i_1}, \ F_{i_2}, \ ..., \ F_{i_k}$—is less than σ, then the sets $F_{i_1}, \ F_{i_2}, \ ..., \ F_{i_k}$ have a non-empty intersection.

Proof : Let us suppose that the assertion is false. Then, there is a sequence of points

(2) $$a_1, \ a_2, \ ..., \ a_m, \ ...$$

and of sub-systems

(3) $$S_1, \ S_2, \ ..., \ S_m, \ ...$$

of the system of sets S such that a_m has a distance less than $1/m$ from all sets of the system S_m, while the intersection of the sets of the system S_m is empty. Since there are only finitely many different sub-systems of the finite system of sets S, there are, in particular among the S_i, only finitely many different systems of sets, so that at least one of them—say S_1—appears in the sequence (3) infinitely often. Consequently, after replacing (2) by a subsequence if necessary, we have the following situation: there is a fixed sub-system

$$S_1 = (F_{i_1}, F_{i_2}, ..., F_{i_k})$$

of S and a convergent sequence of points

(4) $$a_1, \ a_2, \ ..., \ a_m, \ ...$$

with the property that the sets F_{i_h}, $h = 1, 2, ..., k$, have an empty intersection, while, on the other hand, the distance from a_m to each F_{i_h} is less than $1/m$; however, this is impossible, because, under these circumstances the limit point a_ω of the convergent sequence (4) must belong to all sets of the system S_1. q.e.d.

36. For the second lemma we make the following simple observation. Let P be a polyhedron, K a simplicial decomposition of P, and K_1 a subdivision of K. If we let each vertex b of K_1 correspond to the center of a barycentric star containing b, then (by the remark made at the beginning of Sec. **32**) the vertex b is mapped onto a vertex of the simplex of K

containing b; so that, this procedure gives rise to a simplicial map f of K_1 into K. The mapping f—to which we give the name *canonical displacement of K_1 with respect to K*—satisfies the condition of the third conservation theorem, and, hence, gives as the image of the complex K_1 the whole complex K.[46]

The same conclusion still holds if, instead of f, we consider the following *modified* canonical displacement f': first, we displace the vertex b *a little*—that is, less than $\epsilon = \epsilon(K)$ [see assertion b) in Sec. 32]—and then define the center of the barycentric star containing the image of the displacement as the image point $f'(b)$. Then by the previously mentioned assertion b) it follows immediately that the condition of the third conservation theorem is also fulfilled for the mapping f', and consequently, $f'(K_1) = K$.[47]

37. Now that we have defined the concept of canonical displacement (and that of modified canonical displacement) for each subdivision of the complex K, we introduce the same concept for *each sufficiently fine* (curved) simplicial decomposition Q of the polyhedron P, where now Q is *independent* of the simplicial decomposition K except for the single condition that the diameter of the elements of Q must be smaller than the Lebesgue number of the barycentric covering of the polyhedron P corresponding to K. We consider the following mapping of the complex Q into the complex K: to each vertex b of Q we associate the center of one of those barycentric stars of K which contains the point b. The barycentric stars which contain the different vertices of a simplex y of Q are all at a distance of less than the diameter of y from an arbitrarily chosen vertex of the simplex y; since this diameter is smaller than the Lebesgue number of the barycentric covering, the stars in question have a non-empty intersection, and *their centers are thus vertices of one simplex of K*. Our vertex correspondence thus actually defines a simplicial mapping g of Q into K; this mapping g we call a *canonical displacement* of Q with respect to K.

38. Now we are in possession of all the lemmas which are necessary for a very short proof of the invariance theorem for the Betti numbers. Let P

---

[46] The analogous assertion also holds with respect to every algebraic subcomplex of K_1 (or K); if C is an algebraic subcomplex of K, and C_1 a subdivision of C induced by K_1, then the conditions of the third conservation theorem are again fulfilled and we have $f(C_1) = C$.

[47] Similarly, $f'(C_1) = C$ (see the preceding footnote).

and P' be two homeomorphic polyhedra, and K and K' arbitrary simplicial decompositions of them. We wish to show that the r-dimensional Betti number p of K is equal to the r-dimensional Betti number p' of K'. From symmetry considerations it suffices to prove that $p' \geq p$.

For this purpose we notice first of all, that under the topological mapping t of P' onto P, the complex K' and each subdivision K_1' of K' go over into curved simplicial decompositions of the polyhedron P. If we denote, for the moment, by σ a positive number which is smaller than the Lebesgue number of the barycentric covering of K, and also smaller than the number $\epsilon(K)$ defined in Sec. 32, then one can choose the subdivision K_1' of K' so fine that under the mapping t the simplexes and the barycentric stars of K_1' go over into point sets whose diameters are less than σ. These point sets form the (curved) simplexes and the barycentric stars of the simplicial decomposition Q of P into which t takes the complex K_1'. Now let K_1 be a subdivision of K so fine that the simplexes of K_1 are smaller than the Lebesgue number of the barycentric covering of Q. Then there exists (according to Sec. 37) a canonical displacement g of K_1 with respect to Q; furthermore, let f be a canonical displacement of Q with respect to K (this exists because the simplexes of Q are smaller than the Lebesgue number of the barycentric covering of K). Since, by means of g, every vertex of K_1 is moved to the center of a barycentric star of Q containing it, and, therefore, is displaced by *less than* $\epsilon(K)$, the simplicial mapping $f[g(K_1)]$—written $fg(K_1)$ for short—of the complex K_1 into the complex K is a modified canonical displacement of K_1 with respect to K, under which, by Sec. 36,

$$fg(K_1) = K.$$

Furthermore, if C is an algebraic subcomplex of K and C_1 its subdivision in K_1, then (according to footnote 47)

$$fg(C_1) = C.$$

39.   Now let

$$Z_1, Z_2, ..., Z_p$$

be p linearly independent (in the sense of homology) r-dimensional cycles in K and

$$z_1, z_2, ... z_p$$

their subdivisions in K_1. The cycles

$$g(z_1), \; g(z_2), \; ..., \; g(z_p)$$

are independent in Q, since if U is a subcomplex of Q bounded by a linear combination

$$\sum_i c^i g(z_i),$$

then $f(U)$ will be bounded by

$$\sum_i c^i \, fg(z_i),$$

i.e., $\Sigma \, c^i Z_i$, which, according to the assumed independence of the Z_i in K, implies the vanishing of the coefficients c^i.

Under the topological mapping t, the linearly independent cycles $g(z_i)$ of the complex Q go over into linearly independent cycles of the complex K_1' (indeed, both complexes have the same combinatorial structure), so that there are at least p linearly independent r-dimensional cycles in K_1'. Since we have assumed the equality of the Betti numbers of K' and K_1', it follows, therefore, that $p' \geq p$. q.e.d.

With the same methods (and only slightly more complicated considerations) one could also prove the isomorphism of the Betti groups of K and K'.

40. The proof of the theorem of the invariance of Betti numbers which we have just given, following Alexander and Hopf, is an application of the general method of *approximation of continuous mappings of polyhedra by simplicial mappings*. We wish to say here a few more words about this method. Let f be a continuous mapping of a polyhedron P' into a polyhedron P'', and let the complexes K' and K'' be simplicial decompositions of the polyhedra P' and P'', respectively. Let us consider a subdivision K_1'' of K'' so fine that the simplexes and the barycentric stars of K_1'' are smaller than a prescribed number ϵ; then, we choose the number δ so small that two arbitrary points of P' which are less than δ apart go over by means of f into points of P'' whose separation is less than the Lebesgue number σ of the barycentric covering of K_1''. Now consider a subdivision K_1' of K' whose simplexes are smaller than δ. The images of the vertex frames of K_1' have a diameter $< \sigma$, and their totality can be considered as an abstract complex Q; because of the smallness of the simplexes of Q, one can apply to this complex the procedure of Sec. **37**, i.e., one can map it by means of a canonical displacement g into the complex K_1''. The transition of K_1' to Q and the map g from Q to $g(Q)$ together produce a simplicial mapping f_1 of K_1' into K_1''. This mapping (considered as a mapping from P' into P'') differs from f by less than ϵ (i.e., for every point a of P' the distance between the points $f(a)$ and $f_1(a)$ is less than ϵ). *The mapping f_1 is called a simplicial approximation of the continuous mapping f* (and, indeed, one *of fineness ϵ*).

By means of the mapping f_1 there corresponds to each cycle z of K' (where z is to be regarded as belonging to the subdivision K_1' of K') a cycle $f_1(z)$ of K_1''. Moreover, one can easily convince oneself that if $z_1 \sim z_2$ in K' then it follows that $f_1(z_1) \sim f_1(z_2)$ in K_1'', so that to a class of homologous cycles of K' there corresponds a class of homologous cycles of K_1''. In other words, there is a mapping of the Betti groups of K' into the corresponding Betti groups of K_1''; since this mapping preserves the group operation (addition), it is, in the language of algebra, a homomorphism. But there also exists a uniquely determined isomorphism [48] between the Betti groups of K_1'' and K'', so that as a result, we obtain a homomorphic mapping of the Betti groups of K' into the corresponding groups of K''.

Consequently, we have the following fundamental theorem (first formulated by Hopf):

A continuous mapping f of a polyhedron P' into a polyhedron P'' induces a uniquely determined homomorphic mapping of all the Betti groups of the simplicial decomposition K' of P' into the corresponding groups of the simplicial decomposition K'' of P''.[49]

If the continuous mapping f is one-to-one (therefore, topological) it induces an isomorphic mapping of the Betti groups of P' onto the corresponding Betti groups of P''.[50]

By this theorem a good part of the topological theory of continuous

---

[48] Resulting from the canonical displacement of K_1'' with respect to K''.

[49] Because of the isomorphism between Betti groups of the same dimension of different simplicial decompositions of a polyhedron, one can speak simply of the Betti groups of P' or P''.

[50] The proof of this last assertion must be omitted here. Our considerations up to now contain all the elements of the proof; its execution is, therefore, left to the reader. The reader should observe, however, that an arbitrarily fine simplicial approximation to a topological mapping need by no means be a one-to-one mapping.

mappings of polyhedra (in particular of manifolds) is reduced to the investigation of the homomorphisms induced by these mappings, and thus to considerations of a purely algebraic nature. In particular, one arrives at far reaching results concerning the fixed points of a continuous mapping of a polyhedron onto itself.[51]

41. We shall close our treatment of topological invariance theorems with a few remarks about the general concept of dimension which are closely related to the ideas involved in the above invariance proofs. Our previous considerations have paved the way for the following definition.

A continuous mapping f of a closed set F of R^n onto a set F' lying in the same R^n is called an ϵ-*transformation* of the set F (into the set F') if every point a of F is at a distance less than ϵ from its image point $f(a)$.

We shall now prove the following theorem, which to a large extent justifies the general concept of dimension from the intuitive geometrical standpoint, and allows the connection between set-theoretic concepts and the methods of polyhedral topology to be more easily and simply understood than do the brief and, for many tastes, too abstract remarks concerning projection spectra (Sec. **34**).

Transformation theorem. For each $\epsilon > 0$, every r-dimensional set F can be mapped continuously onto an r-dimensional polyhedron by means of an ϵ-transformation; on the other hand, for sufficiently small ϵ there is no ϵ-transformation of F into a polyhedron whose dimension is at most $r - 1$.

The proof is based on the following remark. If

$$(1) \qquad\qquad F_1, \ F_2, \ ..., \ F_s$$

is an ϵ-covering of F, then the nerve of the system of sets (1) is defined first as an abstract complex: to each set $F_i\,(1 \leq i \leq s)$ let there correspond a "vertex" a_i and consider a system of vertices

$$a_{i_0}, \ a_{i_1}, \ ..., \ a_{i_r}$$

---

[51] I mean here principally the Lefschetz-Hopf fixed-point formula which completely determines (and, indeed, expresses by algebraic invariants of the above homomorphism) the so-called *algebraic* number of fixed points of the given continuous mapping (in which every fixed point is to be counted with a definite multiplicity which can be positive, negative, or zero). Concerning this, see Hopf, *Nachr. Ges. Wiss. Gottingen* (1928), pp. 127-136, and *Math. Z.*, Vol. 29 (1929), pp. 493-525.

as the vertex frame of a simplex [of the nerve K of (1)] if and only if the sets F_{i_0}, F_{i_1}, ..., F_{i_r} have a non-empty intersection. However, one can realize this abstract complex *geometrically* if one chooses for a_i a point of F_i itself, or a point from an arbitrarily prescribed neighborhood of F_i, and then allows the vertex frame of the nerve to be spanned by ordinary geometrical simplexes. This construction is always possible, and yields as the nerve of the system of sets (1) an ordinary geometrical polyhedral complex provided the coordinate space R^n in which F lies is of high enough dimension;[52] but this condition can always be satisfied because one can, if need be, imbed the R^n in which F lies in a coordinate space of higher dimension.

42.   In any case, we now assume that a_i is at a distance less than ϵ from F_i for each i, and prove the following two lemmas.

Lemma I.   If K is a geometrically realized nerve of the ϵ-covering (1) of F, then every complex Q whose vertices belong to F, and whose simplexes are smaller than the Lebesgue number σ of the covering (1), goes over into a subcomplex of K by means of a 2ϵ-displacement of its vertices.

Indeed, associate to each vertex b of Q as the point $f(b)$ one of those vertices a_i of K which correspond to the sets F_i containing the point b. Thereby, a simplicial mapping f, of Q into K, is determined; since the distance between a and $f(a)$ is clearly less than 2ϵ, our lemma is proved.

Lemma II.   The conclusion of lemma I also holds (with 3ϵ in place of 2ϵ) if the vertices of Q do not necessarily belong to F but if one knows that they lie at a distance of less than $1/3\,\sigma$ from F, and that the diameters of the simplexes of Q do not exceed the number $1/3\,\sigma$.

In order to reduce this lemma to the preceding one, it is only necessary to transform the vertices of Q into points of F by means of a $1/3\,\sigma$-displacement.

We now decompose the R^n into simplexes which are smaller than $1/3\,\sigma$, and denote by Q the complex which consists of all of those simplexes

---

[52] Indeed, it is sufficient that $n \geq 2r + 1$: with this condition, if one chooses the points a_i in the sets F_i, or in arbitrary neighborhoods of these sets, but so that no $r + 1$ of the points a_i lie in an $(r - 1)$-dimensional hyperplane of R^n, then an elementary consideration shows that our construction is "free of singularities," i.e., that the simplexes do not degenerate and have as their intersections the common faces determined by their common vertices.

which contain points of F in their interiors or on their boundaries; then apply to this complex the lemma just proved. This gives us that:

A sufficiently small polyhedral neighborhood Q of F is transformed by means of a 2ϵ-transformation into a polyhedron P, consisting of simplexes of K.

Since F was r-dimensional and the dimension of the nerve of a system of sets is always 1 less than the order of the system of sets, we may assume that P is at most r-dimensional. From the fact that a certain neighborhood of F is transformed onto the polyhedron P by the 2ϵ-transformation in question, it follows that F itself will be mapped onto a proper or improper subset of P (i.e., *in* P).

Thus we have proved: For every $\epsilon > 0$, F can be mapped onto a subset Φ of an r-dimensional polyhedron by an ϵ-transformation.

We now consider a simplicial decomposition K of P whose elements are smaller than ϵ. Since Φ is closed, there exists—if $\Phi \neq P$—an r-dimensional simplex x^r of K which contains a homothetic simplex x_0^r free of points of Φ. If one now allows the domain $x^r - x_0^r$ which lies between the boundaries of x^r and x_0^r to contract to the boundary of x^r, then all the points of Φ contained in x^r will be transported to the boundary of the simplex x^r, and the points of the set Φ will be "swept out" of the interior of the simplex x^r. By a finite number of repetitions of this "sweeping out" procedure, all r-dimensional simplexes which do not belong to Φ will be freed of points of this set. One continues the process with $(r - 1)$-dimensional simplexes, and so on. The procedure ends with a polyhedron composed of simplexes (of different dimensions) of K. Φ is mapped *onto* this polyhedron by means of a continuous deformation in which no point of Φ leaves that simplex of K to which it originally belonged; consequently, every point of Φ is displaced by less than ϵ. Hence, the whole passage from F to P is a 2ϵ-transformation of the set F, so that the first half of our theorem is proved.

In order to prove the second half, we prove the following more general statement: there exists a fixed number $\epsilon(F) > 0$ such that the r-dimensional set F can be mapped by an $\epsilon(F)$-transformation into no set whose dimension is at most $r - 1$.

We assume that there is no such $\epsilon(F)$. Then, for every $\epsilon > 0$ there exists a set F_ϵ of dimension at most $r - 1$ into which F can be mapped by means of an ϵ-transformation. Consider an ϵ-covering of the set F_ϵ

$$(2) \qquad\qquad F_1^\epsilon, \ F_2^\epsilon, \ ..., \ F_s^\epsilon$$

of order $\leq r$, and denote by F_i the set of all points of F which are mapped into F_i^ϵ by our transformation. Clearly, the sets F_i form a 3ϵ-covering of F of the same order as (2), therefore of order $\leq r$. Since this holds for all ϵ, we must have dim $F \leq r - 1$, which contradicts our assumption. With this, the transformation theorem is completely proved.

43. *Remark.* If the closed set F of R^n has no interior points, then for every ϵ it may be ϵ-transformed into a polyhedron of dimension at most $n - 1$: it suffices to decompose the R^n into ϵ-simplexes and to "sweep out" each n-dimensional simplex of this decomposition. A set without interior points is thus at most $(n - 1)$-dimensional. Since, on the other hand, a closed set of R^n which possesses interior points is necessarily n-dimensional (indeed, it contains n-dimensional simplexes!), we have proved:

A closed subset of R^n is n-dimensional if and only if it contains interior points.

With this we close our sketchy remarks on the topological invariance theorems and the general concept of dimension—the reader will find a detailed presentation of the theories dealing with these concepts in the literature given in footnote 4 and above all in the books of Herr Hopf and the author which have been mentioned previously.

44. *Examples of Betti groups.* 1. The one-dimensional Betti group of the circle as well as of the plane annulus is the infinite cyclic group; that of the lemniscate is the group of all linear forms $u\zeta_1 + v\zeta_2$ (with integral u and v).

2. The one-dimensional Betti number of a $(p + 1)$-fold connected plane region equals p (see Fig. 13, $p = 2$).

3. A closed orientable surface of genus p has for its one-dimensional Betti group the group of all linear forms

$$\sum_{i=1}^{p} u^i \xi_i + \sum_{i=1}^{p} v^i \eta_i \text{ (with integral } u^i \text{ and } v^i);$$

here one takes as generators ξ_i and η_i the homology classes of the $2p$ canonical closed curves.[53]

---

[53] See for example: Hilbert and Cohn-Vossen, p. 264, p. 265, and p. 284 [German edition—A.E.F.].

The non-orientable closed surfaces are distinguished by the presence of a non-vanishing one-dimensional *torsion group*, where by torsion group (of any dimension) we mean the subgroup of the full Betti group con-

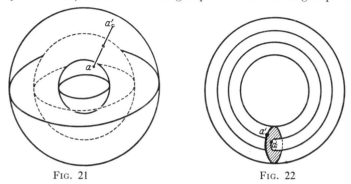

FIG. 21   FIG. 22

sisting of all elements of finite order. The one-dimensional Betti number of a non-orientable surface of genus p is $p - 1$.

The two-dimensional Betti number of a closed surface equals 1 or 0 according as the surface is orientable or not. The analogous assertion also holds for the n-dimensional Betti number of an n-dimensional closed manifold.

4. Let P be a spherical shell (Fig. 21), and Q be the region enclosed between two coaxial torus surfaces (Fig. 22). The one-dimensional Betti number of P is 0, the one-dimensional Betti number of Q is 2, while the two-dimensional Betti numbers of P and Q have the value 1.

5. One can choose as generators of the one-dimensional Betti group of the three-dimensional torus (Sec. 11) the homology classes of the three cycles z_1^1, z_2^1, z_3^1 which are obtained from the three axes of the cube by identifying the opposite sides (Fig. 23). As generators of the two-dimensional Betti group, we can use the homology classes of the three tori into which the three squares through the center and parallel to the sides are transformed under identification. Therefore, the two Betti groups are isomorphic to one another: each has three independent generators, hence, three is both the one- and two-dimensional Betti number of the manifold.

6. For the one- as well as the two-dimensional Betti group of the manifold $S^2 \times S^1$ (see Sec. 11) we have the infinite cyclic group (the corresponding Betti numbers are therefore equal to 1). As z_0^1 choose the cycle (Fig. 21) which arises from the line segment aa' under the identifica-

tion of the two spherical surfaces, and as z_0^2, any sphere which is con-centric with the two spheres S^2 and s^2 and lies between them.

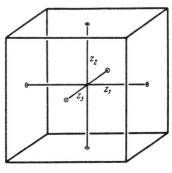

FIG. 23

It is no accident that in the last two examples the one- and two-dimen-sional Betti numbers of the three-dimensional manifolds in question are equal to one another; indeed, we have the more general theorem, known as the *Poincaré duality theorem*, which says that in an n-dimensional closed orientable manifold, the r- and the $(n - r)$-dimensional Betti numbers are equal, for $0 \leq r \leq n$. The basic idea of the proof can be discerned in the above examples: it is the fact that one can choose for every cycle z^r which is not ≈ 0 in M^n a cycle z^{n-r} such that the so-called "intersection number" of these cycles is different from zero.

7. The product of the projective plane with the circle (Sec. 11) is a non-orientable three-dimensional manifold M^3. It can be represented as a solid torus in which one identifies, on each meridian circle, diametrically opposite pairs of points. The one-dimensional Betti number of M^3 is 1 (every one-dimensional cycle is homologous to a multiple of the circle which goes around through the center of the solid torus); the two-dimensional Betti group is finite and has order 2; therefore, it coincides with the torsion group[54] (the torus with the aforementioned identification indeed does not bound, but is a boundary divisor of order 2). Here again there is a general law: the $(n - 1)$-dimensional torsion group of a closed

---

[54] The r-dimensional torsion group $T_r(K)$ of a complex K is the finite group which consists of all elements of finite order of the Betti group $H_r(K)$. The factor group $H_r(K)/T_r(K)$ is isomorphic to $F_r(K)$. [See footnote 21—A.E.F.].

non-orientable n-dimensional manifold is always a finite group of order 2, while an orientable M^n has no $(n - 1)$-dimensional torsion. One can also see from our example that for non-orientable closed manifolds Poincaré's duality theorem does not hold in general.

45. If we consider the polyhedra mentioned in examples 1, 2, and 3 as polyhedra of three-dimensional space, we notice immediately that both the polyhedron and the region complementary to it in R^3 have the same one-dimensional Betti numbers (Figs. 24, 25). This can be seen most easily if one chooses as generators of the group $H_1(P)$ the homology classes of the cycles z_1^1 and z_2^1, respectively, x^1 and y^1, and as the generators of the group $H_1(R^3 - P)$, the homology classes of the cycles Z_1^1 and Z_2^1, respectively, X^1 and Y^1. This remarkable fact is a special case of one of the most important theorems of all topology, the *Alexander duality theorem : the r-dimensional Betti number of an arbitrary polyhedron lying in R^n is equal to the $(n - r - 1)$-dimensional Betti number of its complementary region $R^n - P$, (for $0 < r < n - 1$).*

The proof of Alexander's duality theorem is based on the fact that for every z^r not ≈ 0 in P, there exists a z^{n-r-1} in $R^n - P$ which is linked with it—an assertion whose intuitive sense is made sufficiently clear by figures 24 and 25.[55] This fact also holds for $r = n - 1$ (since pairs of

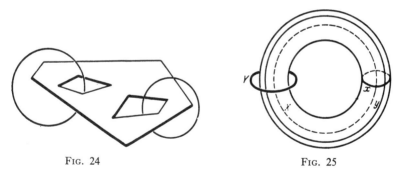

FIG. 24   FIG. 25

---

[55] Concerning the duality theorems of Poincaré and Alexander and the inter-section and linking theories which are closely connected with them, see (in addition to the books by Veblen and Lefschetz): Brouwer, *Amsterd. Proc.*, Vol. 15 (1912) pp. 113-122; Alexander, *Trans. Amer. Math. Soc.*, Vol. 23 (1922) pp. 333-349; Lefschetz, *Trans. Amer. Math. Soc.*, Vol. 28 (1926) pp. 1-49; Van Kampen, *Die kombinatorische Topologie und die Dualitätssätze*, Diss., Leiden, 1929; Pontrjagin, *Math. Ann.*, Vol. 105 (1931) pp. 165-205.

points which are separated by the $(n - 1)$-cycle concerned appear as zero-dimensional linked cycles; see Sec. 2, especially Fig. 1). From these considerations the theorem easily follows that the number of regions into which a polyhedron decomposes R^n is 1 larger than the $(n - 1)$-dimensional Betti number of the polyhedron—a theorem which contains the n-dimensional Jordan theorem as a special case. Both this decomposition theorem and the Alexander duality theorem hold for curved polyhedra.

46. I have intentionally placed at the center of this presentation those topological theorems and questions which are based upon the concepts of the algebraic complex and its boundary: first, because today this branch of topology—as no other—lies before us in such clarity that it is worthy of the attention of the widest mathematical circles; second, because since the work of Poincaré it is assuming an increasingly more prominent position within topology. Indeed, it has turned out that *a larger and larger part of topology is governed by the concept of homology.* This holds true especially for the theory of continuous mappings of manifolds, which in recent years—principally through the work of Lefschetz and Hopf—has shown a significant advance; to a large extent this advance has been made possible by the reduction of a series of important questions to the algebraic investigation of the homomorphisms of the Betti groups induced by continuous mappings (see Sec. 40).[56] Recently, the development of set-theoretic topology, especially that of dimension theory, has taken a similar turn; it is now known that the concepts of cycle, boundary, Betti groups, etc., hold not only for polyhedra, but also can be generalized to include the case of arbitrary closed sets. Naturally, the circumstances here are much more complicated, but even in these general investigations we have now advanced so far that we are at the beginning of a systematic and entirely geometrically oriented theory (in the sense of the program outlined in Sec. 4) of the most general structures of space, a theory which has its own significant problems and its own difficulties. This theory is also based principally on the concept of homology.[57]

Finally, the part of topology which is concerned with the concepts of cycle and homology is the part on which the applications of topology depend

---

[56] In addition to the works of Hopf (footnote 8) and Lefschetz (footnote 55) cited previously, see Hopf, *J. f. Math.*, Vol. 165 (1931) pp. 225-236.

[57] See the works of the author given at the end of footnote 4.

almost exclusively; the first applications to differential equations, mechanics, and algebraic geometry lead back to Poincaré himself. In the last few years these applications have been increasing almost daily. It suffices here to mention, for example, the reduction of numerous analytical existence proofs to topological fixed point theorems, the founding of enumerative geometry by Van der Waerden, the pioneering work of Lefschetz in the field of algebraic geometry, the investigations of Birkhoff, Morse and others in the calculus of variations in the large, and numerous differential geometrical investigations of various authors, etc.[58] One may say, without exaggeration: *anyone who wishes to learn topology with an interest in its applications must begin with the Betti groups*, because today, just as in the time of Poincaré, most of the threads which lead from topology to the rest of mathematics and bind topological theories together into a recognizable whole lead through this point.

---

[58] A rather complete bibliography will be found at the end of the book by Lefschetz, already mentioned many times before.

INDEX